T0224041

# Advances in Intelligent Systems and Computing

## Volume 568

**Series editor**

Janusz Kacprzyk, Polish Academy of Sciences, Warsaw, Poland
e-mail: kacprzyk@ibspan.waw.pl

*About this Series*

The series "Advances in Intelligent Systems and Computing" contains publications on theory, applications, and design methods of Intelligent Systems and Intelligent Computing. Virtually all disciplines such as engineering, natural sciences, computer and information science, ICT, economics, business, e-commerce, environment, healthcare, life science are covered. The list of topics spans all the areas of modern intelligent systems and computing.

The publications within "Advances in Intelligent Systems and Computing" are primarily textbooks and proceedings of important conferences, symposia and congresses. They cover significant recent developments in the field, both of a foundational and applicable character. An important characteristic feature of the series is the short publication time and world-wide distribution. This permits a rapid and broad dissemination of research results.

*Advisory Board*

Chairman

Nikhil R. Pal, Indian Statistical Institute, Kolkata, India
e-mail: nikhil@isical.ac.in

Members

Rafael Bello Perez, Universidad Central "Marta Abreu" de Las Villas, Santa Clara, Cuba
e-mail: rbellop@uclv.edu.cu

Emilio S. Corchado, University of Salamanca, Salamanca, Spain
e-mail: escorchado@usal.es

Hani Hagras, University of Essex, Colchester, UK
e-mail: hani@essex.ac.uk

László T. Kóczy, Széchenyi István University, Győr, Hungary
e-mail: koczy@sze.hu

Vladik Kreinovich, University of Texas at El Paso, El Paso, USA
e-mail: vladik@utep.edu

Chin-Teng Lin, National Chiao Tung University, Hsinchu, Taiwan
e-mail: ctlin@mail.nctu.edu.tw

Jie Lu, University of Technology, Sydney, Australia
e-mail: Jie.Lu@uts.edu.au

Patricia Melin, Tijuana Institute of Technology, Tijuana, Mexico
e-mail: epmelin@hafsamx.org

Nadia Nedjah, State University of Rio de Janeiro, Rio de Janeiro, Brazil
e-mail: nadia@eng.uerj.br

Ngoc Thanh Nguyen, Wroclaw University of Technology, Wroclaw, Poland
e-mail: Ngoc-Thanh.Nguyen@pwr.edu.pl

Jun Wang, The Chinese University of Hong Kong, Shatin, Hong Kong
e-mail: jwang@mae.cuhk.edu.hk

More information about this series at http://www.springer.com/series/11156

Rituparna Chaki · Khalid Saeed · Agostino Cortesi
Nabendu Chaki
Editors

# Advanced Computing
# and Systems for Security

## Volume Four

Springer

*Editors*
Rituparna Chaki
A.K. Choudhury School of Information
    Technology
University of Calcutta
Kolkata, West Bengal
India

Khalid Saeed
Faculty of Computer Science
Bialystok University of Technology
Białystok
Poland

Agostino Cortesi
DAIS—Università Ca' Foscari
Mestre, Venice
Italy

Nabendu Chaki
Department of Computer Science and
    Engineering
University of Calcutta
Kolkata, West Bengal
India

ISSN 2194-5357          ISSN 2194-5365   (electronic)
Advances in Intelligent Systems and Computing
ISBN 978-981-10-3390-2      ISBN 978-981-10-3391-9   (eBook)
DOI 10.1007/978-981-10-3391-9

Library of Congress Control Number: 2016960766

This Springer imprint is published by Springer Nature
The registered company is Springer Nature Singapore Pte Ltd.
The registered company address is: 152 Beach Road, #21-01/04 Gateway East, Singapore 189721, Singapore

# Preface

The Third International Doctoral Symposium on Applied Computation and Security Systems (ACSS 2016) took place during August 12–14, 2016 in Kolkata, India organized by University of Calcutta in collaboration with NIT, Patna. The Ca Foscari University of Venice, Italy and Bialystok University of Technology, Poland were the international collaborators for ACSS 2016.

As the symposium turned to its third consecutive year, it has been interesting to note the kind of interests it has created among the aspiring doctoral scholars and the academic fraternity at large. The uniqueness of being a symposium created specifically for Ph.D. scholars and their doctoral research has already established ACSS as an inimitable event in this part of the globe. The expertise of the eminent members of the program committee helped in pointing out the pros and cons of the research works being discussed during the symposium. Even during presentation of each paper, the respective Session Chair(s) had responded to their responsibilities of penning down a feedback for more improvements of the paper. The final version of the papers thus goes through a second level of modification as per the session chair's comments, before being included in this post-conference book.

Team ACSS is always on the look-out for latest research topics, and in a bid to allow such works, we always include them to our previous interest areas. ACSS 2016 had Android Security as the chosen new interest area, thus inviting researchers working in the domains of data\analytics, cloud and service management, security systems, high performance computing, algorithms, image processing, pattern recognition.

The sincere effort of the program committee members, coupled with indexing initiatives from Springer, has drawn a large number of high-quality submissions from scholars all over India and abroad. A thorough double-blind review process has been carried out by the PC members and by external reviewers. While reviewing the papers, reviewers mainly looked at the novelty of the contributions, at the technical content, at the organization and at the clarity of presentation. The entire process of paper submission, review and acceptance process was done electronically. Due to the sincere efforts of the Program Committee and the Organizing Committee members, the symposium resulted in a suite of strong technical paper presentations followed by effective discussions and suggestions for improvement for each researcher.

The Technical Program Committee for the symposium selected only 21 papers for publication out of 64 submissions after peer review. Later 5 more papers were chosen for presenting in the symposium after the authors submitted enhanced versions and those were reviewed again in a 2-tier pre-symposium review process.

We would like to take this opportunity to thank all the members of the Program Committee and the external reviewers for their excellent and time-bound review works. We thank all the sponsors who have come forward towards organization of this symposium. These include Tata Consultancy Services (TCS), Springer Nature, ACM India, M/s Neitec, Poland, and M/s Fujitsu, Inc., India. We appreciate the initiative and support from Mr. Aninda Bose and his colleagues in Springer Nature for their strong support towards publishing this post-symposium book in the series "Advances in Intelligent Systems and Computing". Last but not least, we thank all the authors without whom the symposium would have not reached this standard.

On behalf of the editorial team of ACSS 2016, we sincerely hope that ACSS 2016 and the works discussed in the symposium will be beneficial to all its readers and motivate them towards even better works.

| | |
|---|---|
| Kolkata, India | Rituparna Chaki |
| Białystok, Poland | Khalid Saeed |
| Mestre, Venice, Italy | Agostino Cortesi |
| Kolkata, India | Nabendu Chaki |

# Contents

# About the Editors

**Rituparna Chaki** is Professor of Information Technology in the University of Calcutta, India. She received her Ph.D. Degree from Jadavpur University in India in 2003. Before this she completed B.Tech. and M.Tech. in Computer Science and Engineering from University of Calcutta in 1995 and 1997, respectively. She has served as a System Executive in the Ministry of Steel, Government of India for 9 years, before joining the academics in 2005 as a Reader of Computer Science and Engineering in the West Bengal University of Technology, India. She has been with the University of Calcutta since 2013.

Her areas of research include optical networks, sensor networks, mobile ad hoc networks, Internet of Things, data mining, etc. She has nearly 100 publications to her credit. Dr. Chaki has also served in the program committees of different international conferences. She has been a regular Visiting Professor in the AGH University of Science and Technology, Poland for last few years. Rituparna has co-authored couple of books published by CRC Press, USA.

**Khalid Saeed** is Full Professor in the Faculty of Computer Science, Bialystok University of Technology, Bialystok, Poland. He received the B.Sc. degree in Electrical and Electronics Engineering in 1976 from Baghdad University in 1976, the M.Sc. and Ph.D. degrees from Wroclaw University of Technology, in Poland in 1978 and 1981, respectively. He received his D.Sc. degree (Habilitation) in Computer Science from Polish Academy of Sciences in Warsaw in 2007. He was a visiting professor of Computer Science with Bialystok University of Technology, where he is now working as a full professor. He was with AGH University of Science and Technology during 2008–2014. He is also working as Professor with the Faculty of Mathematics and Information Sciences in Warsaw University of Technology. His areas of interest include biometrics, image analysis and processing and computer information Systems. He has published more than 220 publications, edited 28 books, journals and conference proceedings, 10 text and reference books. He has supervised more than 130 M.Sc. and 16 Ph.D. theses. He gave more than 40 invited lectures and keynotes in different conferences and universities in Europe, China, India, South Korea and Japan on biometrics, image analysis and processing. He has received more than 20 academic awards. Khalid Saeed is a member of more than 20 editorial boards of international journals and conferences. He is an IEEE Senior Member and has been selected as IEEE Distinguished Speaker for 2011–2016. Khalid Saeed is the Editor-in-Chief of International Journal of Biometrics with Inderscience Publishers.

**Agostino Cortesi, Ph.D.** is Full Professor of Computer Science at Ca' Foscari University, Venice, Italy. He served as Dean of the Computer Science studies, as Department Chair, and as Vice-Rector for quality assessment and institutional affairs.

His main research interests concern programming languages theory, software engineering, and static analysis techniques, with particular emphasis on security applications. He published more than 110 papers in high-level international journals and proceedings of international conferences. His h-index is 16 according to Scopus, and 24 according to Google Scholar. Agostino served several times as member (or chair) of program committees of international conferences (e.g., SAS, VMCAI, CSF, CISIM, ACM SAC) and he's in the editorial boards of the journals "Computer Languages, Systems and Structures" and "Journal of Universal Computer Science". Currently, he holds the chairs of "Software Engineering", "Program Analysis and Verification", "Computer Networks and Information Systems" and "Data Programming".

**Nabendu Chaki** is Professor in the Department of Computer Science and Engineering, University of Calcutta, Kolkata, India. Dr. Chaki did his first graduation in Physics from the legendary Presidency College in Kolkata and then in Computer Science and Engineering from the University of Calcutta. He completed Ph.D. in 2000 from Jadavpur University, India. He shares six international patents including four U.S. patents with his students. Professor Chaki has been quite active in developing international standards for software engineering and cloud computing as a member of Global Directory (GD) member for ISO-IEC. Besides editing more than 25 book volumes, Nabendu has authored six text and research books and has more than 150 Scopus Indexed research papers in journals and international conferences. His areas of research interests include distributed systems, image processing and software engineering. Dr. Chaki has served as a Research Faculty in the Ph.D. program in Software Engineering in U.S. Naval Postgraduate School, Monterey, CA. He is a visiting faculty member for many universities in India and abroad. Besides being in the editorial board for several international journals, he has also served in the committees of over 50 international conferences. Professor Chaki is the founder Chair of ACM Professional Chapter in Kolkata.

# Part I
# Cloud and Service Management

# A Review on Energy Efficient Resource Management Strategies for Cloud

Srimoyee Bhattacherjee, Sunirmal Khatua and Sarbani Roy

**Abstract** Green computing has acquired significant amount of importance in the present research scenario as with mammoth advancements in the field of Information and Communications Technology (ICT), energy consumption has increased manifold over the last decade. With development in services provided by cloud providers, a large amount of energy is being consumed by the cloud data centers distributed all over the world. These data centers have a major contribution to the carbon footprints that is being generated, which in turn is detrimental for the environment. This paper studies the current researches that are being conducted to build energy efficient cloud networks and explores strategies that can be adopted to reduce the energy consumption of cloud data centers, thus opening new avenues towards building a 'green' cloud network.

**Keywords** Energy efficient · Green · Cloud · Data center · Resource · Scheduling · Load balancing · Migration

## 1 Introduction

Improvement of system performance has been of utmost priority to the designers of computing systems and the industry. As stated in Moore's Law, the number of transistors in dense integrated circuits has doubled approximately every two years.

S. Bhattacherjee (✉) · S. Roy
Department of Computer Science and Engineering,
Jadavpur University, Kolkata, India
e-mail: b.srimoyee@gmail.com

S. Roy
e-mail: sarbani.roy@cse.jdvu.ac.in

S. Khatua
Department of Computer Science and Engineering,
University of Calcutta, Kolkata, India
e-mail: skhatuacomp@caluniv.ac.in

© Springer Nature Singapore Pte Ltd. 2017
R. Chaki et al. (eds.), *Advanced Computing and Systems for Security*,
Advances in Intelligent Systems and Computing 568,
DOI 10.1007/978-981-10-3391-9_1

This is the reason behind the steady growth of performance of computing systems [1]. This has led to the increase of the total power drawn by computing. The degree of the similar problem is large in case of large-scale computing systems, such as clusters and data centers. The power consumption of data centers has been found to have doubled between 2000 and 2005 in a study conducted by the Lawrence Berkeley National Laboratory. It is now nearly 1.2% of American electricity consumption. Companies now spend as much as 10% of their technology budgets on energy. This inclination can only promote the energy costs of a server during its lifespan to exceed its cost of hardware [2, 3].

Now-a-days the buzz is "Cloud computing" everywhere around us. Cloud Computing has been defined as *a model for enabling ubiquitous, convenient, on-demand network access to a shared pool of configurable computing resources (e.g., networks, servers, storage, applications, and services) that can be rapidly provisioned and released with minimal management effort or service provider interaction* by the National Institute of Standards and Technology (NIST) [4]. The metaphor "cloud" is actually being used to refer to large capacity data centers and servers which are providing various kinds of services to its consumers in an efficient way. These high end computing systems have resulted to rising energy consumption of computing systems. Eventually, further performance growth has to be restricted due to increasing electricity bills and carbon emissions.

Cloud infrastructure providers are establishing Data Centers in various locations all over the world. As these data centers consume huge amount of electric power, these Data Centers have become expensive to operate [5, 6]. This has various adverse effects on our society as it is increasing the power bills, along with contributing to global warming due to high carbon emission. It is the ICT sector that is presently responsible for about 2% of global greenhouse gas emissions. Efficient usage of resources, avoiding utilization of excess and unnecessary resources, and reducing the carbon footprint is an open challenge that can be addressed to minimize energy consumption of Data Centers.

With immense increase in Internet usage with Google and Facebook leading the way, there has been significant rise in energy consumption. In 1995, less than 1% of the world population had access to the Internet which has increased to 40% presently. Each year is experiencing rise in the number of Internet users. As of 2013, in every 60 s, 204 million email messages are exchanged [7], Google experiences 5 million searches [8], Facebook witnesses 1.8 million "likes" [9], Twitter gets 350,000 tweets [10], products worth $272,000 is sold via e-commerce websites like Amazon [7] and 15,000 tracks get downloaded from iTunes [11]. All these online activities are delivered through data centers and this growth in usage of the Internet is only likely to rise with time. Hence, the demands on data centers will grow which will facilitate the increase of energy consumption.

A report published by International Data Corporation (IDC) [12] also predicts IT energy costs will stay on a rising curve if nothing is done towards power and energy consumptions. Thus, finding effective approaches to cut energy costs is a paramount issue for organizations to protect their business, as well as the global environment.

In this work, Sect. 2 gives an overview of energy efficient cloud computing, followed by a brief discussion on existing resource management techniques in Sect. 3. Section 4 presents a detailed study of existing energy efficient resource management approaches and conclusion and future scope for this work is discussed in Sect. 5.

## 2 Energy Efficient Cloud Computing: An Overview

In a typical cloud computing environment, a service request reaches the cloud broker from the cloud consumer. The broker decides to which cloud service provider (CSP) the request is to be assigned. In a cloud network, a federation consists of one or more CSP. A federation is formed to match business needs and for better performance. It is the concept of interconnecting the cloud computing environments of two or more CSPs for the purpose of load balancing and better performance. Any CSP can be a part of more than one federation, combining with various sets of different CSPs to service a request in a better way. Each CSP consists of one or more data centers (DC). Each of these DCs consists of one or more physical machines (PM) or resources. A service request is first assigned to the most suitable federation available where its constituent CSPs process it in a distributed fashion. The CSPs send the request to one of its DCs which in turn sends it to one of its PM which actually executes the service.

A typical cloud computing environment can be put forward in a tuple form as shown below,

$$F = \left\{ CSP_P | CSP_P = \left\{ DC_d^p | DC_d^p = PM_m^{pd} \right\} \right\}$$
$$PM = <processor\_type,\ speed,\ memory,\ storage,\ QoS,\ power>$$
$$SLA = <QoS,\ Usage\_Cost>$$

where, F is any Cloud Federation, CSP is Cloud Service Provider, DC is Data Center, PM is Physical Machine, and QoS is the Quality of Service provided by the PM. The SLA or the Service Level Agreement chiefly comprises of all the QoS metrics and the usage cost which should be strictly maintained.

For building a green cloud environment, the significant amount of carbon footprints generated by the cloud datacenters has to be minimized. Any cloud datacenter consists of a number of physical machines or resources. These physical machines at any point in time can be in any of the three modes, namely, *active*, *idle* or *terminated*.

In a cloud environment, all the physical machines of all the datacenters cannot be in active mode at any particular time instance. Some may be in the active mode while others may be in the idle or sleep modes. Therefore, for a time period T, the total energy consumption of a typical cloud environment, $E_{C_{(T)}}$ is given by,

$$\forall p = 1, 2, \ldots, |CSP|, \forall d = 1, 2, \ldots, |DC|, \forall m = 1, 2, \ldots, |PM|,$$

$$E_{C_{(T)}} = E_{(T)}^1 + E_{(T)}^2 + \cdots + E_{(T)}^p,$$

where, $E_{(T)}^p = E_{(T)}^{p1} + E_{(T)}^{p2} + \cdots + E_{(T)}^{pd}$,

where, $E_{(T)}^{pd} = E_{1_{(T)}}^{pd} + E_{2_{(T)}}^{pd} + \cdots + E_{m_{(T)}}^{pd}$,

therefore, $E_{C_{(T)}} = \sum_{t=1}^{T} \sum_{p=1}^{|CSP|} \sum_{d=1}^{|DC_p|} \sum_{m=1}^{|PM_{pd}|} E_{m_{(t)}}^{pd}$

where, $E_{m_{(t)}}^{pd}$ is the energy consumed by the $m$-th PM of the $d$-th DC of the $p$-th CSP at time $t$, which will be dependent on the mode of the physical machine at that time instance.

Keeping in view the current research scenario, it is obvious that conservation of energy is a very crucial aspect that has direct impact on our environment. As cloud computing is getting popular with time, building energy efficient cloud data centers that generate less carbon foot print is one of the most important challenges to the researchers and the industry.

## 3 Existing Resource Management Techniques

It has been observed that efficient resource management is one of the key steps towards achieving energy efficient in cloud networks. Efficient resource scheduling, load balancing, migration and offloading techniques can be used for managing cloud resources in a smarter way. A brief study of such existing techniques will be presented in this section.

### 3.1 Resource Scheduling

Many scheduling algorithms have been developed for allocation of resources in a cloud network. A list of existing algorithms and their corresponding scheduling parameters have been listed below:

(i) A Compromised-Time-Cost Scheduling Algorithm [13]—Cost and time
(ii) SHEFT Workflow Scheduling Algorithm [14]—Execution time and reliability
(iii) Improved cost-based algorithm for task scheduling [15]—Cost and performance
(iv) Multiple QoS constrained Scheduling Strategy of Multi-workflows [16]—Scheduling success rate, cost, time and make span

(v)  A Particle–Swarm Optimization based Heuristic for Scheduling [17]—
     Resource Utilization and time
(vi)  Innovative transaction intensive cost-constraint scheduling algorithm [18]—
     Execution Cost and time

It is observed that the most common scheduling parameters are cost, performance and time. Cost for executing in a certain data center can increase if the data
center consumes more energy while execution. Again, if a task requires more time
for execution when allocated to a certain data center, it eventually means more
amount of energy consumption. It is desirable to execute a task on a data center
which consumes less energy, simultaneously assuring that the performance will not
degrade. A tradeoff has to be maintained between performance i.e. QoS and the
energy consumption. Hence 'energy consumed' by a data center should also be
considered as a scheduling parameter while scheduling resources in a cloud
environment.

## 3.2  Load Balancing

The method of load balancing in cloud is applied across different data centers to
guarantee the network availability by minimizing use of computer hardware,
software failures and mitigating recourse limitations [19]. The load balancers
provide two important needs. They are, promotion of availability of cloud resources
and promotion of performance. The performance metrics are directly linked to
energy consumption of data centers. Hence, efficient load balancing schemes are
absolutely necessary to make the cloud networks energy efficient.

It has also been identified in [20] that the objective of load balancing is to reduce
the resource usage which will further decrease energy consumption and carbon
emission rate that is the need of the hour for a cloud network. A few of the
well-known existing load balancing techniques are VectorDot [21], CLBVM [22],
Event-driven [23], CARTON [24], LBVS [25], Compare and Balance [26],
Scheduling strategy on LB of VM resources [27], Task Scheduling based on LB
[28], Honeybee Foraging Behavior [29], Biased Random Sampling [29], Active
Clustering [29], Join-Idle-Queue [30], ACCLB [31], OLB and LBMM [32],
Decentralized content aware [33], Server-based LB for Internet distributed services
[34], and Lock-free multiprocessing solution for LB [35].

It has been observed that all these existing load balancing techniques in clouds,
consider various factors like performance, resource utilization, response time,
migration time, fault tolerance, scalability, throughput and associated overhead. But
to build a green cloud network, the load balancing schemes should also consider
energy consumption for making the load balancing schemes energy aware.

### 3.3 Migration

Migration in cloud is also a very important technique in case a fault appears in the server where a request is being executed currently.

When a virtual machine (VM) is to be migrated from a faulty server to another available server, many servers might be available on which the migrating VM can be allocated at that time instance. Keeping social welfare in mind, the VM should be migrated to such a server which will consume least energy among all the servers available at that instant to host the VM. As migration of VMs is a very common event in a cloud network, energy efficient migration schemes will have a major contribution in building a green cloud network.

Migration of VMs in cloud environment has been a topic of research interest for quite some time. An ideal example in such respect is Windows Azure Fabric [36]. It has a structure like that of a weave, composed of servers and load balancers (referred to as nodes), and power, Ethernet and serial communications (referred to as edges). It has a Fabric Controller that manages a service node through a built-in service called Azure Fabric Controller Agent. If there is a fault, the Controller either reboots the server or migrates services from the current server to other healthy servers. In [37], migration of virtual machines has been spotted as one of the key step towards optimum resource allocation in a cloud computing environment. Heuristics have been proposed in [37] for choosing VMs to migrate, which has been discussed in detail in Sect. 4. The technique of live or on-line migration is being used for efficient resource allocation in cloud computing [37, 38]. According to [39], in cloud computing, VMs may be migrated for statistical multiplexing or dynamically changing communication schemes to attain high bandwidth for tightly coupled hosts or to achieve variable heat distribution and power availability in the data center. A novel approach named EnaCloud has been proposed in [40] which enable application live migration to decrease the number of active machines, in order to save energy. Again, minimizing the number of migrations is an interesting issue that can be addressed in the context of live migration [41]. Presents the minimization of migrations policy in order to achieve an energy efficient cloud network.

From the discussions in Sect. 3, it is clear that resource handling and management in a cloud environment is of utmost importance to reduce energy consumption. Resource scheduling, load balancing and migration are various techniques to efficiently handle cloud resources that help in building of a 'green' cloud environment.

The studies and researches discussed give an idea about how the various resource management schemes play vital roles in conservation of energy. Thus the development of such research should be promoted.

## 4 Energy Efficient Resource Management: A Research Direction

Extensive research has identified efficient resource management to be one of the most important key steps towards building Green Cloud Computing architecture.

Energy efficient resource management in cloud computing refers to the management of cloud resources considering energy consumption as the decision maker. The VM allocation and migration schemes should be designed in such a way so that the overall energy consumption of the cloud environment is minimized.

Any cloud architecture should be designed in a way such that there is a specific unit which looks after the energy aspect of the environment. Such architecture is given in [42] which has a specialized "Green Resource Manager". It has the 'Green Negotiator' which finalizes the SLA metrics between the cloud provider and consumer, with the consumers or brokers depending on the consumers' QoS needs. It also looks after the energy saving schemes. There is also an Energy Monitor in this architecture which is in charge of observing and determining which physical machines to be kept power on/off.

An energy efficient resource management system for virtualized Cloud data centers has been proposed in [43]. Here energy savings have been achieved by continuous consolidation of VMs based on the current utilization of resources, architecture of the virtual network established between VMs and thermal state of computing nodes.

A similar work has been presented in [44] where it has been said that data centers consume enormous amounts of electrical power which results in high costs required for operating and carbon dioxide emissions. Hence an efficient resource management policy for virtualized Cloud data centers has been proposed. The main idea behind this is continuous consolidation of VMs influencing live migration and switching off idle nodes to reduce power consumption, while maintaining the required QoS.

The power and energy models discussed in [41] clearly shows that power consumption by CPU, disk storage and network interfaces contribute to the total power consumption by computing nodes in data centers. Compared to other system resources, it has been observed that the maximum amount of energy is consumed by the CPU. Studies have also revealed that on an average, an inactive server consumes around 70% of the power consumed by the server running at full CPU speed. This leads to the introduction of the technique of switching idle servers off which reduces total power consumption. The power model defined in this work is as follows:

$$P(u) = k*P_{max} + (1 - k)*P_{max}*u \tag{1}$$

where $P_{max}$ is the maximum power consumed by a fully utilized server, $k$ is the fraction of power consumed by the inactive server and $u$ is the CPU utilization. The utilization of CPU changes over time due to variable workload. Thus, $u(t)$ stands for

CPU utilization which is a function of time. Therefore, total energy (E) consumption by a physical server can be calculated by applying integration on the power consumption function over a period of time as represented in Eq. (2).

$$E = \int_t P(u(t)) \tag{2}$$

## 4.1 A Detailed Study

Amongst all the related literatures, relevant algorithms have been found to be discussed in [41]. In [41], Modified Best Fit Decreasing (MBFD) algorithm which is an energy aware algorithm for VM placement on physical machines in a cloud computing environment has been presented. The well-known bin packing problem has been used to form the MBFD algorithm which finds out upon placement on which host, a particular virtual machine consumes least energy among all the available hosts and places it accordingly. To calculate the power and energy consumed by a host, MBFD uses the power and energy models stated in Eqs. (1) and (2) respectively.

As discussed in [41], the optimization problem is carried out in two steps:

(i)  The virtual machines to be migrated are selected;
(ii) the selected virtual machines are placed on the new hosts using MBFD algorithm.

For VM migration, three threshold based methods have been presented in [41] which are:

(i)   The Minimization of Migrations policy (MM)—The minimum number of VMs is selected by this policy which should be migrated from a host to lessen the CPU utilization in case of violation of the upper threshold of the CPU utilization.
(ii)  The Highest Potential Growth policy (HPG)—VMs that have the least usage of the CPU compared to the CPU capacity stated by the VM parameters is migrated using this policy. Minimizing the potential increase of the host's utilization and prevent SLA violation in case of violation of the upper threshold of the CPU utilization is the aim of this policy.
(iii) The Random Choice policy (RC)—This policy is dependent on the random selection of VMs needed to lower the CPU utilization of a host when the CPU utilization of the host exceeds the upper utilization threshold.

Analysis of the performance of each of these migration policies and the change in energy consumption, average SLA violations, number of VM migrations w.r.t the lower utilization threshold has been discussed in [41]. These policies also seem to reduce the energy consumption in data centers.

MBFD algorithm has been designed based on the well-known bin packing problem. Here, the physical machines or hosts are considered as the bins and the VMs are considered as objects of different volumes which have different specifications and must be fitted (allocated) into the physical machines (bins), based on energy consumptions. The idea of the Best Fit Decreasing bin packing has been used to formulate MBFD.

Again, in [45], a few more single threshold based migration policies have been discussed, namely, MC (Maximum Correlation), MU (Minimum Utilization), and MMT (Minimum Migration Time).

### 4.1.1 An Observation

All the above mentioned policies in [41, 45] have been implemented using *Cloudsim 3.0.3*, using workload from PlanetLab, on a homogeneous set of VMs. On application of the threshold based migration policies (MM, RC, MC, MU, MMT), it has been observed that in most of the cases, the MM policy does not consume the least energy, as desired, compared to that of the other policies, as shown in Fig. 1. Moreover, there were a few cases where the RC policy generated less energy than the MM policy, though the number of VM migrations was large in case of RC policy. A random policy ideally should produce worst results but in this case, there were exceptional scenarios.

It could be inferred that the MM policy does not produce best results in all kind of scenarios.

To make the simulation environment more realistic, we generated a heterogeneous set of VMs and placed them on the physical machines using MBFD, following which we implemented the threshold based migration policies and observed that the RC policy continues to yield better results in some of the cases. Thus the efficient behavior of the MM policy is found to be case specific which is not desirable.

It was observed that the authors of [41, 45] have implemented all the migration policies in *Cloudsim 3.0.3* by only considering the energy required for VM

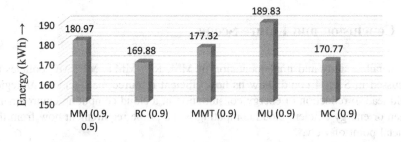

**Fig. 1** Energy consumptions (in kWh) for MM, RC, MMT, MU, MC policies for homogeneous VMs

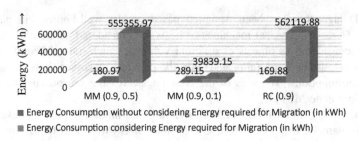

Fig. 2  Energy consumptions (in kWh) for MM and RC policies for homogeneous VMs

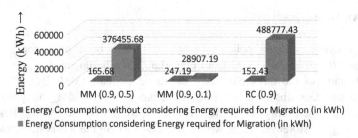

Fig. 3  Energy consumptions (in kWh) for MM and RC policies for heterogeneous VMs

placement i.e. the energy required to migrate a VM from one host to another has not been considered. Thus, we included the energy incurred for VM migration in the MM and RC migration policies for both homogeneous and heterogeneous set of hosts and found that the MM policy produces the best results in all cases. As the number of VM migrations in case of RC policy is large compared to that of MM policy, the energy consumed by MM policy is always the least, even after considering the energy consumed in migration. Thus we can conclude that the energy consumed in migrating contributes to a great extent in the overall energy consumption of a cloud datacenter. Figures 2 and 3 show that the MM policy consumes least amount of energy when the energy required for migrating VMs is considered for calculating the overall energy consumption.

## 5   Conclusion and Future Scope

The detailed study and implementation of MM, RC, MMT, MU, MC policies as discussed in Sect. 4 could show us how efficient resource management strategies could lead to reduction of energy consumption in a cloud computing environment. Green or energy efficient cloud computing is the prime requirement now from the societal point of view.

In future, this work can be extended to compare the MM policy with other energy efficient migration policies. The policies mentioned here, other than MM, were not designed keeping energy consumption in mind. It can also be checked if there is any other heuristic that can be used to minimize the overall energy consumption of a cloud network.

We can also concentrate on the decision making factor which decides whether a VM is to be migrated or not. Prediction models can be used to predict the future behaviors of the physical machines and the virtual machines by recording and analyzing their past behaviors. Based on the predicted results, it can be decided whether to initiate a migration or not so that the overall energy consumption is reduced.

The cloud algorithms that have been studied focus either on optimizing the energy consumption or the overall cost. While formulating an energy efficient cloud algorithm, the behavior of the cost function is not considered, and vice versa. Designing a VM placement and migration algorithm for cloud which will optimize both the energy consumption and the overall cost is also another scope of work in this area. The concept of multiobjective optimization is to be brought into address this scenario.

**Acknowledgements** The research of the first author is supported by DST—Inspire Fellowship vide Reference No. DST/INSPIRE Fellowship/IF140873.

# References

1. Moore, G.E.: Cramming more components onto integrated circuits. IEEE Solid-State Circuits Soc. Newsl. **11**(5), 33–35 (2006)
2. Beloglazov, A., Buyya, R., Lee, Y.C., Zomaya, A.: A taxonomy and survey of energy-efficient data centers and cloud computing systems. Adv. Comput. **82**(2), 47–111 (2011)
3. Barroso, L.: The price of performance. ACM Queue **3**(7), 53 (2005)
4. Mell, P., Grance, T.: The NIST definition of cloud computing. National Institute of Standards and Technology, vol. 53, no. 6 (2009)
5. Buyya, R., Pandey, S., Vecchiola, C.: Cloudbus toolkit for market-oriented cloud computing. In: Cloud Computing, pp. 24–44 (2009)
6. Buyya, R., Calheiros, R.N., Li, X.: Autonomic cloud computing: open challenges and architectural elements. In: 2012 IEEE 3rd International Conference on Emerging Applications of Information Technology (EAIT), pp. 3–10 (2012)
7. Domo.com. http://www.domo.com/learn/infographic-data-never-sleeps
8. Statisticbrain.com. http://www.statisticbrain.com/google-searches/
9. Gizmodo.com. http://gizmodo.com/5937143/what-facebook-deals-with-everyday-27-billion-likes-300-million-photos-uploaded-and-500-terabytes-of-data
10. Internetlivestats.com. http://www.internetlivestats.com/twitter-statistics/
11. Billboard.com. http://www.billboard.com/biz/articles/news/1538108/itunes-crosses-25-billion-songs-sold-now-sells-21-million-songs-a-day
12. IDC. Transforming the datacenter: consolidation, pervasive virtualization and energy optimization (2009)
13. Liu, K., Yang, Y., Chen, J., Liu, X., Yuan, D., Jin, H.: A Compromised-time-cost scheduling algorithm in SwinDeW-C for instance-intensive cost-constrained workflows on cloud computing platform. Int. J. High Perform. Comput. Appl. **24**(4), 445–456 (2010)

14. Lin, C., Lu, S.: Scheduling scientific workflows elastically for cloud computing. In: 2011 IEEE International Conference on Cloud Computing (CLOUD), pp. 746 –747 (2011)
15. Selvarani, S., Sadhasivam, G.S.: Improved cost-based algorithm for task scheduling in cloud computing. In: 2010 IEEE international conference on computational intelligence and computing research (ICCIC), pp. 1–5 (2010)
16. Xu, M., Cui, L., Wang, H., Bi, Y.: A multiple QoS constrained scheduling strategy of multiple workflows for cloud computing. In: 2009 IEEE international symposium on parallel and distributed processing with applications, pp. 629–634 (2009)
17. Pandey, S., Wu, L., Guru, S.M., Buyya, R.: A particle swarm optimization-based heuristic for scheduling workflow applications in cloud computing environments. In: 2010 IEEE 24th International Conference on Advanced Information Networking and Applications (AINA), pp. 400–407 (2010)
18. Yang, Y., Liu, K., Chen, J., Liu, X., Yuan, D., Jin, H.: An algorithm in SwinDeW-C for scheduling transaction-intensive cost-constrained cloud workflows. In: Proceedings of the 2008 4th International Conference on e-Science, pp. 374–375 (2008)
19. Chaczko, Z., Mahadevan, V., Aslanzadeh, S., Mcdermid, C.: Availability and load balancing in cloud computing. In: 2011 international conference on computer and software modeling (IPCSIT) (2011)
20. Kansal, N.J., Chana, I.: Cloud load balancing techniques: a step towards green computing. Int. J. Comput. Sci. 9(1), 2238–2246 (2012)
21. Singh, A., Korupolu, M., Mohapatra, D.: Server-storage virtualization: integration and load balancing in data centers. In: Proceedings of the 2008 ACM/IEEE conference on Supercomputing, p. 53 (2008)
22. Bhadani, A., Chaudhary, S.: Performance evaluation of web servers using central load balancing policy over virtual machines on cloud. In: Proceedings of the 2010 3rd Annual ACM Bangalore Conference, p. 16 (2010)
23. Nae, V., Prodan, R., Fahringer, T.: Cost-efficient hosting and load balancing of massively multiplayer online games. In: Proceedings of the 2010 11th IEEE/ACM International Conference on Grid Computing, pp. 9–17 (2010)
24. Stanojevic, R., Shorten, R.: Load balancing vs. distributed rate limiting: an unifying framework for cloud control. In: IEEE International Conference on Communications (ICC), pp. 1–6 (2009)
25. Liu, H., Liu, S., Meng, X., Yang, C., Zhang, Y.: Lbvs: A load balancing strategy for virtual storage. In: 2010 IEEE International Conference on Service Sciences (ICSS), pp. 257–262 (2010)
26. Zhao, Y., Huang, W.: Adaptive distributed load balancing algorithm based on live migration of virtual machines in cloud. In: Proceedings of the 2009 5th International Joint Conference on INC, IMS and IDC, pp. 170–175 (2009)
27. Hu, J., Gu, J., Sun, G., Zhao, T.: A scheduling strategy on load balancing of virtual machine resources in cloud computing environment. In: 2010 IEEE 3rd International Symposium on Parallel Architectures, Algorithms and Programming (PAAP), pp. 89–96 (2010)
28. Fang, Y., Wang, F., Ge, J.: Existing load balancing techniques in cloud computing: a systematic review. Lect. Notes Comput. Sci. 3(1), 271–277 (2010)
29. Randles, M., Lamb, D., Taleb-Bendiab, A.: A comparative study into distributed load balancing algorithms for cloud computing. In: 2010 IEEE 24th International Conference on Advanced Information Networking and Applications Workshops (WAINA), pp. 551–556 (2010)
30. Lua, Y., Xiea, Q., Kliotb, G., Gellerb, A., Larusb, J.R., Greenber, A.: Join-Idle-Queue: a novel load balancing algorithm for dynamically scalable web services. Perform. Eval. 68(11), 1056–1071 (2011)
31. Zhang, Z., Zhang, X.: A load balancing mechanism based on ant colony and complex network theory in open cloud computing federation. In: 2010 IEEE 2nd International Conference on Industrial Mechatronics and Automation (ICIMA), pp. 240–243 (2010)

32. Wang, S., Yan, K., Liao, W., Wang, S.: Towards a load balancing in a three-level cloud computing network. In: 2010 IEEE 3rd International Conference on Computer Science and Information Technology (ICCSIT), pp. 108–113 (2010)

33. Mehta, H., Kanungo, P., Chandwani, M.: Decentralized content aware load balancing algorithm for distributed computing environments. In: Proceedings of the ACM International Conference Workshop on Emerging Trends in Technology, pp. 370–375 (2011)

34. Nakai, A.M., Madeira, E., Buzato, L.E.: Load balancing for internet distributed services using limited redirection rates. In: 2011 IEEE 5th Latin-American Symposium on Dependable Computing (LADC), pp. 156–165 (2011)

35. Liu, X., Pan, L., Wang, C.J., Xie, J.Y.: A lock-free solution for load balancing in multi-core environment. In: 2011 IEEE 3rd International Workshop on Intelligent Systems and Applications (ISA), pp. 1–4 (2011)

36. Windows Azure Platform. http://www.microsoft.com/azure/

37. Buyya, R., Beloglazov, A., Abawajy, J.: Energy-efficient management of data center resources for cloud computing: a vision, architectural elements, and open challenges. In: Proceedings of the 2010 International Conference on Parallel and Distributed Processing Techniques and Applications (2010)

38. Berl, A., Gelenbe, E., Girolamo, M.D., Giuliani, G., Meer, H.D., Dang, M.Q., Pentikousis, K.: Energy-efficient cloud computing. Comput. J. 53(7), 1045–1051 (2010)

39. Zhang, Q., Cheng, L., Boutaba, R.: Cloud computing: state-of-the-art and research challenges. J. Internet Serv. Appl. 1(1), 7–18 (2010)

40. Li, B., Li, J., Huai, J., Wo, T., Li, Q., Zhong, L.: EnaCloud: an energy-saving application live placement approach for cloud computing environments. In: 2009 IEEE International Conference on Cloud Computing (CLOUD), pp. 17–24 (2009)

41. Beloglazov, A., Abawajy, J., Buyya, R.: Energy-aware resource allocation heuristics for efficient management of data centers for cloud computing. Future Gener. Comput. Syst. 28(5), 755–768 (2012)

42. Buyya, R., Vecchiola, C., Selvi, S.T.: Mastering Cloud Computing. Tata McGraw Hill Education (2013)

43. Beloglazov, A., Buyya, R.: Energy efficient resource management in virtualized cloud data centers. In: Proceedings of the 2010 10th IEEE/ACM International Conference on Cluster, Cloud and Grid Computing, pp. 826–831 (2010)

44. Beloglazov, A., Buyya, R.: Energy efficient allocation of virtual machines in cloud data centers. In: 2010 IEEE/ACM 10th International Conference on Cluster, Cloud and Grid Computing (CCGrid), pp. 577–578 (2010)

45. Beloglazov, A., Buyya, R.: Optimal online deterministic algorithms and adaptive heuristics for energy and performance efficient dynamic consolidation of virtual machines in cloud data centers. Concurr. Comput.: Pract. Exp. 24(13), 1397–1420 (2012)

# Modeling and Analysis of Enterprise Cloud Bus Using a Petri Net Based Approach

**Gitosree Khan, Sabnam Sengupta and Anirban Sarkar**

**Abstract** Agent based Cloud computing technology has become a predominant domain nowadays due to rapid increase in enterprise service applications that are built from various services offered by different cloud service providers. In our previous work, we have proposed an abstraction layer of SaaS architecture for Inter-cloud environment, called Enterprise Cloud Bus System (ECBS) and modeled it using Unified Modeling Language (UML). In this work, we have proposed a conceptual architecture of Multi-agent based Enterprise Cloud Bus System (ECBS) and modelled its dynamics using Petri Net (ECBP) based approach. The proposed approach will be beneficial for analyzing low level Petri Nets of any cloud based models and has the advantage of being able to model dynamic facets of the Multi-agent based cloud system that can handle multiple agent operations and its execution logics. Further, the proposed ECBP model is capable to analyze the behavioral facets of Enterprise Cloud Bus System like, fairness, boundedness, liveliness, safeness, etc. in a dead lock—free environment.

**Keywords** Cloud computing · Enterprise Cloud Bus System · Behavioral analysis · Enterprise Cloud Bus Petri Net · PIPE Petri Net tool

G. Khan (✉) · S. Sengupta
B. P. Poddar Institute of Management & Technology, Kólkata, India
e-mail: khan.gitosree@gmail.com

S. Sengupta
e-mail: sabnam_sg@yahoo.com

A. Sarkar
National Institute of Technology, Durgapur, India
e-mail: sarkar.anirban@gmail.com

© Springer Nature Singapore Pte Ltd. 2017    17
R. Chaki et al. (eds.), *Advanced Computing and Systems for Security*,
Advances in Intelligent Systems and Computing 568,
DOI 10.1007/978-981-10-3391-9_2

# 1  Introduction

Cloud computing technology delivers on demand service over the network as per user's requirement. In recent days, due to rapid increase of cloud and its services as well as the growing interest of enterprise towards service oriented computing [1–4] discovery and selection leads to the problem of flexible and scalable access to computing resources. Lots of researchers [5, 6] discuss over the architectural design of cloud computing and its applications. Many of them focus on the vision, challenges and architectural elements [7] of Inter-cloud environments. The work in [8, 9] covered on dynamic provisioning, deployment and adaptation of dynamic Multi-cloud behaviour systems. The author in [10] proposes a new approach for dynamic autonomous resource management in cloud computing.

In recent days, challenges and research directions of agent oriented software engineering has been explained in [11]. Formalization and analysis of Multi-agent based software system has been proposed in [12]. There are few works [13–15] based on measurement and validation of complexity metrics of Multi-agent based system architecture. Further the author in [16] models the Multi-agent system dynamics using graph semantic based approach. Moreover, modeling and design of agent based hybrid and flexible Multi-cloud architecture [17, 18] has emerged as one of the most challenging domains in cloud computing. Thus, as our previous work are based upon such MAS based Multi-cloud architecture. In [19], we have proposed a Multi-Agent based abstraction layered architecture, called Enterprise Cloud Bus (ECB) and modeled its structural components using UML 2.0. [20], to make the cloud based system more reliable and robust. Few of our earlier work are based on service registration and discovery mechanism in ECBS [21–23] which helps to identify services during run time. Further, in [24, 25] scheduling of services and its performance analysis in ECBS has been proposed. However, UML cannot be used for automatic analyses and simulation of Inter-cloud architecture, because of its Semi-formal nature.

Since, the UML modeling lacks to exhibit the dynamism of internal behavior of the system. PIPE [26, 27] is a Platform Independent Petri Net Editor tool that helps to analyze low level Petri Nets and has the advantage of being able to model dynamic behavior of the Multi-agent based cloud system and handle Multiple agent operations and its execution logics. In [28–32] the author proposed a Petri net based approach for modeling and analysis of agent oriented system. In such a domain, few of the research work [33] are done on modeling of Inter-cloud architecture using Petri Net tool.

This work enhances the modeling and analysis of Multi-cloud architecture (*ECBS*) using Petri Net based approach (ECBP), in order to address the issues of the dynamic facets of the system. The proposed method helps to analyze and verified the behavioral properties of the system like, reachability, safeness, boundedness, liveness, etc. This kind of formal approach is helpful for effective design of *MAS* based Inter-cloud architecture by providing maximum utilization of resources on demand through scheduling agent, automation of resources and dynamic provisioning of all types of heterogeneous cloud services. Further, a case study has been explained based on the proposed approach.

# 2 Enterprise Cloud Bus System (ECBS)

Enterprise Cloud Bus System (ECBS) describes a high level abstraction layer of SaaS architecture in Inter-cloud environment, where different clouds interacts and collaborates through cloud agent from various locations in order to publish and/or subscribe their services. The detailed set of building blocks in the context of *ECBS* has been described in Fig. 1.

## 2.1 Formal Definition of ECBS

In this section, we have discussed about the formal definition of *ECBS* framework. The CloudBus (*CB*) is a set of agents and components and is structurally defined as:

$$CB \rightarrow CLIENT \wedge PA \wedge CUDDI \wedge ESB \wedge CESB \wedge CA \wedge HUDDI \wedge SA \wedge MAPPER \wedge LOGGER \wedge RES \tag{1}$$

A Multi-cloud environment *Multi-CloudEnv* is that where components of *CB* will work using the following four tuples. It can be formally defined as,

$$Multi - CloudEnv = \{Res, Actor, CB, Relation\} \tag{2}$$

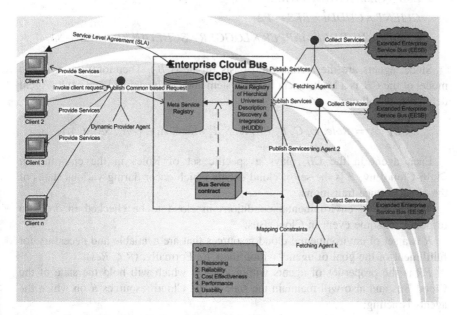

**Fig. 1** Enterprise Cloud Bus System Framework (ECBS)

In the given Cloud environment, *Res* is the set of cloud resource, *Actors* are clients of the cloud environment, *CB* are the set of autonomous cloud bus entities with pre specified functions and *Relation* is the set of semantic association and interactions among the cloud bus entities.

## 2.2 Conceptualization of ECBS in MAS Architecture

A conceptual architecture of ECBS deals with high level representation of the agents and components present inside the cloud architecture. The concept of MAS definition in the proposed architecture is used from [15, 16]. Agent based system are the de facto paradigm to handle the dynamicity of Multi cloud architecture like ECBS.

The dynamicity of $CB_i$ in the environment *Multi-CloudEnv* are handled using three agents {*PA, SA, CA*} and other components relevant to single cloud architecture as described in Fig. 1.

Formally, the dynamic model of any *CloudBus* ($CB_i$) can be defined as,

$$CB_i = \{CA_i, PA_i, SA_i\} \tag{3}$$

Each of the agents within the CloudBus ($CB_i$) will be invoked if the following conditions hold by different agents;

$$ESB \wedge CESB \wedge HUDDI \wedge RES \rightarrow CA_i$$

$$CLIENT \wedge CUDDI \wedge RES \rightarrow PA_i$$

$$CUDDI \wedge HUDDI \wedge MAPPER \wedge LOGGER \wedge SCHEDULER \wedge RES \rightarrow SA_i$$

Since, agents are the architectural basis of the ($CB_i$). Therefore, the dynamic model of the $CB_i$ is a Multi-agent based system and can be defined as a Multi-Agent definition as follows:

$$CB_i \cdot A_i = [Role, E, C, R, PR, K, S, I] \text{ where, } A_i \in \{PA_i, SA_i, CA_i\}.$$

Each agent in the $CB_i$ plays a specific set of Roles in the environment Multi-CloudEnv; $E$ is the set of cloud events which occur during various states of the cloud service transitions.

$C$ is a set of environmental conditions in cloud to be checked in order to response on some event in Cloud Bus;

$R$ is a set of environmental cloud resources that are available and necessary for fulfillment of the goal of agents within the $CB_i$. Formally, ($R \subset Res$);

$PR$ is the properties of agents within the $CB_i$ which will hold the state of the Cloud Bus and also will maintain the state of the Cloud resources $R$ on which the agents is acting;

*K* is the set of information that forms the main knowledge base. Initially it comprises of the states of available cloud resources that the agents within the $CB_i$ will use to response on some event. The *K* can be update dynamically.

*S* is the set of cloud services that the agents within the $CB_i$ can provide and conceptualize this will determine the capability of the Cloud Bus components;

*I* is a set of interactions between the agents reside inside the $CB_i$.

## 2.3  ECBS Elements in MAS Architecture

The roles, events and related services along with the respective resources, properties and knowledge base of each agents present within the $CB_i$ is summarized in Table 1.

Here, Provider Agent (*PA*), Cloud Agent (*CA*) and Scheduling Agent (*SA*) starts working with the minimal set of knowledge of the environment to render the request, service and resource token as shown in Table 1. Secondly, all the agents

**Table 1**  Role collaboration templates of ECBS

| Role | Event | Service | Cloud bus component |
|------|-------|---------|---------------------|
| Agent: Provider Agent (PA) | | | |
| R0: Request Transmitter R1: Service Provider | E0: Request Transmitted E1: Service Provided | S0: SendRequest S1: ProvideService | CLIENT |
| R2: Request Provider | E2: Request Registered | S2: GetRequest | CUDDI |
| R3: Resource Seeker | E3: Resource used & Released | S3: GetResource S4: ReleseResource | RES |
| Agent: Cloud Agent (CA) | | | |
| R4: Service Invoker | E4: Service Invoked | S5: InvokeService | ESB |
| R5: Service Collector | E5: Service Collected | S6: CollectService | CESB |
| R6: Service Transmitter | E6: Service Registered | S7: PublishService | HUDDI |
| R7: Resource Seeker | E7: Resource used & Released | S8: GetResource S9: ReleseResource | RES |
| Agent: Scheduler Agent (SA) | | | |
| R8: Service Matcher | E8: Service Matched | S10: MatchService | CUDDI |
| R9: Service Seeker | E9: Service Discovered | S11: GetService | HUDDI |
| R10: Service Mapper | E10: Service Mapped | S12: MapService | MAPPER |
| R11: Service Scheduler | E11: Service Scheduled | S13: ScheduleService | SCHEDULER |
| R12: Service Logger | E12: Service Logged | S14: LogService | LOGGER |
| R13: Service Dispatcher | E13: Service Dispatched | S15: DispatchService | CLIENT |
| R14: Resource Seeker | E14: Resource used & Released | S16: GetResource S17:ReleseResource | RES |

will use several properties to hold the state of the resources and the states of itself. Finally, the knowledge base is updated dynamically once the component of CloudBus starts working. With all these and with some defined set of Interactions $I$, the agents will be performing some services like

$S = \{MatchService, GetService, MapService, ScheduleService, LogService,$
$DispatchService, GetResource, ReleaseResource\}.$

## 3 Proposed Enterprise Cloud Bus Petri Net (ECBP)

Petri Nets are a graphical modeling tool that makes a dynamic system more formalized. Therefore, in this section to analyze the behavior of ECBP model we are using a Petri Net tool called PIPE to ensure correctness and performance of the system at design time. The proposed model provides better utilization and automation of cloud resources.

### 3.1 Definition: Enterprise Cloud Bus Petri Net (ECBP)

A Multi-Agent based Enterprise Cloud Bus Petri Net, *ECBP*, is defined by the 5-tuples as follows:

(a) $ECBP = [P, T, \Pi, I, O, H, M0, W]$;
(b) $P = \{P1, P2 ..., Pm\}$ is a finite set of places.
(c) $T = \{T1, T2, Tn\}$ is a finite set of transitions, $S \cap T = \emptyset$.
(d) $I, O, H: T \rightarrow N$ ($N = S \cup T$), are the input, output and inhibition functions.
(e) M0 is the initial marking where, $M0: S \rightarrow IN$ is the initial marking.
(f) $\Pi: T \rightarrow IN$ is the priority function that associates with the lower priorities to timed transitions and higher priorities to immediate transitions. Immediate transitions therefore always have priority over timed ones.
(g) $W$ is a weight function $W: F \rightarrow \{1, 2 ...\}$. $W: T \rightarrow R$ is a function that associates with a real value to the transitions.

The various components of *ECBP* can be mapped to the elements of the conceptual framework as under. The mapping rules are as follows:

A place $P$ in ECBP comprises of set of tokens $t_k$ (Request, Service) belongs to constraints, properties, knowledge and services, roles, interactions of any *ECBS* components. Formally, $P \rightarrow t_k$ where, $t_k \in C \cup K \cup PR \cup K \cup S$. All the events $E$ will be mapped as transitions $T$ of a PN. Formally, $T \rightarrow E$.

A component of *ECBS* will request for a resource. Once a resource will be allocated to a component it will hold the resource until the next transition is fired from that component.

The graphical notation of place and transition are represented as Circle and Bar respectively. In *ECBP* it is important to note that, due to firing of any transition *T*, all the sub components of the place *P* will be affected simultaneously. Hence for any place in *ECBP* one can set the mark as 0 or 1.

## 3.2 ECBP Elements: Places and Transitions

The various elements of the proposed *ECBP* will be $\Sigma = [Colour\ set\ for\ request\ token,\ Colour\ set\ for\ service\ token]$. *C* is the Color function where, $C_r = \{black\ for\ request\ token\}$ and $C_s = \{blue\ for\ service\ token\}$. The color set value of tokens (Request, Service) are marked as $P = 1;\ Q = 2$. The places *P*, transitions *T* and tokens t have been summarized in Table 2 and Table 3 respectively.

**Table 2** Places and transitions with its descriptions based on Table 1

| Places | Component of places | Transitions | Events |
|--------|--------------------|-------------|--------|
| P0 | Client | T0 | E0, E3 |
| P1 | PA | T1 | E2, E3 |
| P2 | CUDDI | T2 | E4, E7 |
| P3 | ESB | T3 | E5, E7 |
| P4 | CESB | T4 | E6, E7 |
| P5 | CA | T5 | E8, E14 |
| P6 | HUDDI | T6 | E9, E14 |
| P7 | SA | T7 | E10, E14 |
| P8 | MAPPER | T8 | E11, E14 |
| P9 | SCHEDULER | T9 | E12, E14 |
| P10 | LOGGER | T10 | E13, E14 |

**Table 3** Token and its descriptions

| Places | Token | Description of tokens | Token parameters | Color set value of token |
|--------|-------|----------------------|------------------|--------------------------|
| P0 | t0 | Request | Sent | 1 |
| P1 | t0 | Request | Provide | 1 |
| P2 | t0 | Request | Register | 1 |
|    | t1 | Service | Register | 2 |
| P3 | t1 | Service | Provide | 2 |
| P4 | t1 | Service | Publish | 2 |
| P5 | t1 | Service | Provide | 2 |
| P6 | t1 | Service | Collect | 2 |
| P7 | t0 | Request | Match | 1 |
|    | t1 | Service | Register | 2 |
| P8 | t1 | Service | Discover | 2 |
| P9 | t1 | Service | Map | 2 |
| P10 | t1 | Service | Schedule | 2 |

# 4   Analysis of ECBS Based on ECBP

Enterprise Cloud Bus Petri Net (*ECBP*) is a suitable tool to model the behaviour of ECBS system. Moreover, several features of dynamic system like, occurrence of finite number of events, deadlock free operations, achievement of goals through firing of events etc. can be analyzed through the analysis of *ECBP* properties like, safeness, boundedness, liveness, reachability etc.

## 4.1   ECBP Based Analysis of ECBS

Figure 2 shows the *ECBP* net of *ECBS* system. The process starts from a place *P0* and after a transition *T1* will reach a place *P1*. The process continues further on and finally arrives at the place *P10*.

The process from *P0* to *P2* for the transition *T0* to *T2* and from *P3* to *P6* for the transition *T3* to *T6* are occur concurrently because dynamically at the same time clients are registering their request through provider agent in *CUDDI* and services are published by the cloud agent in *HUDDI*. Next, the process is interrupted for service matching with the client request and then from place *P6* it will follow the path of places *P3*, *P7*, *P8*, *P9* and *P10* for the transition *T3*, *T7*, *T8*, *T9* and *T10* respectively. If the service is not matched with the required client request the process is cancelled and will be rescheduled.

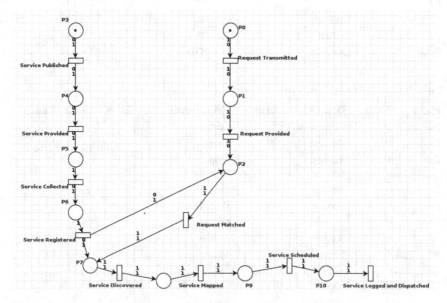

**Fig. 2**  A Petri Net modeling of Enterprise Cloud Bus System using PIPE: ECBP

### 4.1.1 Structural Analysis

The structural analysis of the proposed net is shown in Fig. 3. Here, the marking graph is constructed to analyze certain behavioral aspects of the system. We define a siphon as a set of places $P$ satisfying $S \subseteq P$; a siphon is marked by a marking $m$ if at least one of its places is marked by m. In *ECBP* with a siphon $S$, if $S$ is not marked at the initial marking, then $S$ is not marked at any reachable marking of the proposed system. Similarly, we consider sets of places that never lose all tokens once at least one of their places is marked.

A sufficient condition to siphon is that each transition that removes at least one token from this set also adds a token. This condition is formalized further by the concept of trap. In general, a trap is a set $T$ of places satisfying $T \subseteq T(P)$; a trap is marked by a marking $m$ if at least one of its places is marked at $m$.

### 4.1.2 Performance Analysis

The performance analysis of the proposed model is based on the number of reachable tangible markings which is shown in Fig. 4. The corresponding steady-state distribution of tangible states of the model is obtained as follows:

$$X \pi Q = 0 \, M \in RS \, \pi[M] = 1;$$

where $Q$ is the infinite simal generator matrix, whose diagonal elements are the rates of the exponential distributions associated with the transitions from one state to other, while the elements on the main diagonal describes the sum of the elements of each row equal to zero. $\Pi$ is the equilibrium probability mass function (pmf) over the reachable markings $M$; usually, we write $\pi[M]$ for the steady-state probability of a given marking $M$.

**Fig. 3** Results for structural analysis of ECBS

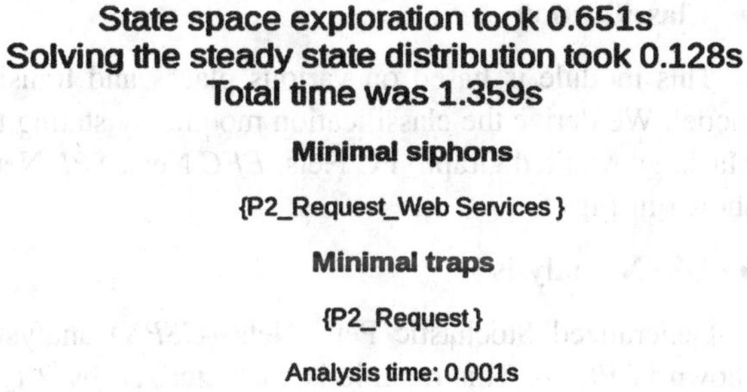

**State space exploration took 0.651s**
**Solving the steady state distribution took 0.128s**
**Total time was 1.359s**

**Minimal siphons**

{P2_Request_Web Services}

**Minimal traps**

{P2_Request}

**Analysis time: 0.001s**

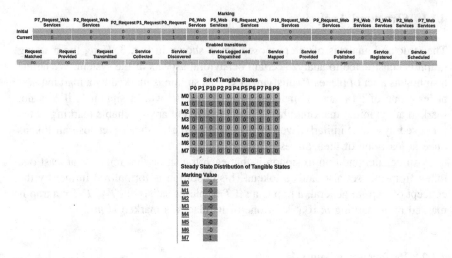

**Fig. 4** Results for performance analysis of ECBS

**Fig. 5** Classification module

| State Machine | false |
|---|---|
| Marked Graph | true |
| Free Choice Net | true |
| Extended Free Choice Net | true |
| Simple Net | true |
| Extended Simple Net | true |

### 4.1.3 Qualitative and Quantitative Analysis Modules

In this section, we describe different types of qualitative and quantitative analysis modules of Multi-Agent based *ECBS* framework using *ECBP*. The following are the analysis modules of *ECBP:*

- **Classification**

  This module is based on various places and transitions connectivity of *ECBP* model. We derive the classification module by stating the classification types; State Machine, Marked Graph, FC Nets, *EFC* Nets, *SPL* Nets, and *ESPL* Nets which are shown in Fig. 5.

- **GSPN analysis**

  Generalized Stochastic Petri Nets (*GSPN*) analysis of *ECBS* using *ECBP* is shown in Fig. 6. This module is characterized by 2 types of transitions:

  (a) Stochastic transitions that is associated with an exponentially distributed firing delay.
  (b) Immediate transitions which are associated with a null firing delay.

| Throughput of Timed Transitions | | Average Number of Tokens on a Place | |
|---|---|---|---|
| Transition | Throughput | Place | Number of Tokens |
| Request Matched | 0 | P0 | 0 |
| Request Provided | 0 | P1 | 0 |
| Request Transmitted | 0 | P10 | 0 |
| Service Collected | 1 | P2 | 0 |
| Service Discovered | 0 | P3 | 0 |
| Service Logged and Dispatched | 0 | P4 | 0 |
| Service Mapped | 0 | P5 | 0 |
| Service Provided | 1 | P6 | 0 |
| Service Published | 1 | P7 | 0 |
| Service Registered | 1 | P8 | 0 |
| Service Scheduled | 0 | P9 | 0 |

| Sojourn times for tangible states | | Token Probability Density | | |
|---|---|---|---|---|
| Marking | Value | | $\mu=0$ | $\mu=1$ |
| M0 | 0.2 | P0 | 1 | 0 |
| M1 | 0.2 | P1 | 1 | 0 |
| M2 | 0.2 | P10 | 1 | 0 |
| M3 | 0.2 | P2 | 1 | 0 |
| M4 | 0.2 | P3 | 1 | 0 |
| M5 | 0.2 | P4 | 1 | 0 |
| M6 | 0.2 | P5 | 1 | 0 |
| M7 | 0.25 | P6 | 1 | 0 |
| | | P7 | 1 | 0 |
| | | P8 | 1 | 0 |
| | | P9 | 1 | 0 |

**Fig. 6** GSPN analysis

Here, from Fig. 6 we calculates the average number of tokens on a place, the token probability density and the throughput of timed transitions by exploring the state space of the given Petri Net model (*ECBP*) and determining the steady state solution of the model.

- **Invariant analysis**

In general the occurrences of transitions refer to the transformation of token over a Petri Net. The total number of token for a given set of places remains unchanged if the pre-set and the post-set of the transitions contain the same number of places of this set. The concept of place and transition invariants formalizes such invariant properties.

A vector $w \in Zm$, $w \neq 0$ is a transition invariant of an *ECBP* with $m$ transitions if $Cw = 0$ where $C$ is the incidence matrix of the given Net. A transition invariant describes a group of transitions that may be fired without affecting the marking of the Net. In this section resulted place invariant, transition invariant and place invariant equations are shown in Fig. 7. A vector $v \in Zn$, $v \neq 0$ is a place invariant of an *ECBP* with $n$ places. If $v T M = v T M0$ for all reachable markings of the *ECBP*.

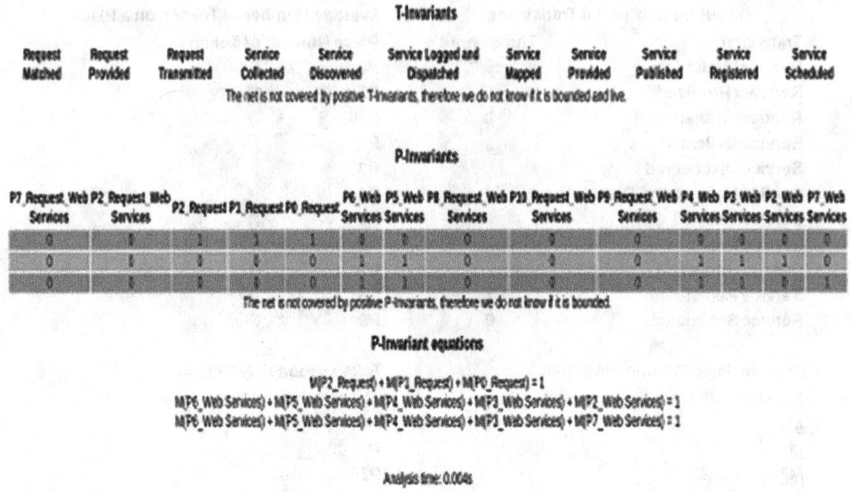

**T-Invariants**

| Request Matched | Request Provided | Request Transmitted | Service Collected | Service Discovered | Service Logged and Dispatched | Service Mapped | Service Provided | Service Published | Service Registered | Service Scheduled |
|---|---|---|---|---|---|---|---|---|---|---|

The net is not covered by positive T-invariants, therefore we do not know if it is bounded and live.

**P-Invariants**

| P7_Request_Web Services | P2_Request_Web Services | P2_Request | P1_Request | P0_Request | P6_Web Services | P5_Web Services | P8_Request_Web Services | P33_Request_Web Services | P9_Request_Web Services | P4_Web Services | P3_Web Services | P2_Web Services | P7_Web Services |
|---|---|---|---|---|---|---|---|---|---|---|---|---|---|
| 0 | 1 | 1 | 1 | 1 | 0 | 0 | 0 | | 1 | 0 | 0 | 0 | 0 |
| 0 | 1 | 1 | 1 | 0 | 1 | 1 | 0 | | 0 | 0 | 1 | 1 | 1 | 0 |
| 0 | 1 | 0 | 1 | 0 | 1 | 1 | 0 | | 0 | 0 | 1 | 1 | 1 | 1 |

The net is not covered by positive P-invariants, therefore we do not know if it is bounded.

**P-Invariant equations**

$$M(P2\_Request) + M(P1\_Request) + M(P0\_Request) = 1$$
$$M(P6\_Web\ Services) + M(P5\_Web\ Services) + M(P4\_Web\ Services) + M(P3\_Web\ Services) + M(P2\_Web\ Services) = 1$$
$$M(P6\_Web\ Services) + M(P5\_Web\ Services) + M(P4\_Web\ Services) + M(P1\_Web\ Services) + M(P7\_Web\ Services) = 1$$

Analysis time: 0.004s

**Fig. 7** Invariant analysis result

- **Incidence and Marking Matrices of ECBP Model**

This module validates the proposed Petri Net model *ECBP* based on the result of the following generated matrices like forward matrix, backward matrix, combined incidence matrices and the inhibition matrix are shown in Fig. 8.

A forward-incidence matrix represents the initial state; backwards-incidence matrix shows the operational state after firing of the set of dynamic events in *ECBS* of combined matrix representing the status of the token at any given instance after the initiation of the token transitions and the inhibition matrix representing the inhibition arcs.

Each of these matrix has been formed using the row constituents *p0_Request*, *p1_Request...p10_Request_Web Services* and the column constituents Request Provided, Request Transmitted,...., Service Scheduled. In the *ECBP* model, among the places *P0, P1, P2, P3, P4, P5, P6, P7, P8, P9, P10* none of them are covered and hence the Net is not covered by *P* invariants. The same is the case for the transitions and the Net is not covered by *T* invariants.

## 4.2 Analysis of Behavioral Properties of ECBP Model

Some of the crucial behavioral properties have been analyzed using the *ECBP* model.

(a) **Safeness**: Any place of a graph is declared as safe, if the number of tokens at that place is either 0 or 1. In the proposed *ECBP*, the set of places [*P1...P10*] represents a combination of 0 (no token) and 1 (token), which implies that if the

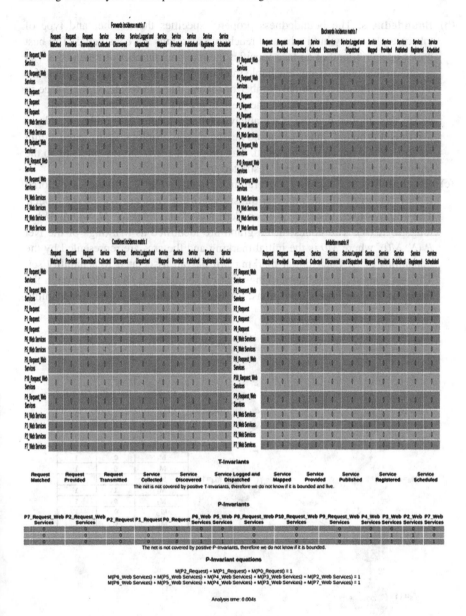

**Fig. 8** Incidence and marking matrices of ECBP model

firing occurs there will be a token at the position bit otherwise no token. Thus it shows each of the places has a maximum token count 1 or 0 and is declare safe. Since, all the places in the *ECBP* are safe then the Net as a whole can be declared safe.

(b) **Boundedness**: The boundedness property specifies the number and type of tokens a place may hold when all reachable markings are considered. The best upper integer bound of a place specifies the maximal number of tokens that can reside on that place in any reachable marking. The best lower integer bound for a place specifies the minimal number of tokens that can reside on that place in any reachable marking. When the best upper and lower integer bounds are equal, this implies that the place always contains a constant number of tokens and thus it is bound. In the proposed net there is no deadlock at any stage within *P1* to *P10* and hence it is bounded. The proposed *ECBP* Net is safe and so the number of tokens in a place is restricted to either 0 or 1.

(c) **Reachability**: Reachability Coverability graph in Fig. 9 provides a pictorial representation of all the possible transitions firing sequences of *ECBP* model. The graph is used to define a given Petri Net *N* and marking *M*, where *M* belongs to *R* (*N*). The nodes of the graph are identified as markings of the Net *R* (*N*, *M0*), where *M0* is the initial marking and the arcs are represented by the transitions of *N*. Each initial marking *M0* has an associated Reachability set. This set consists of all the markings that can be reached from *M0* through the firing of one or more transitions. In our proposed model the reachability

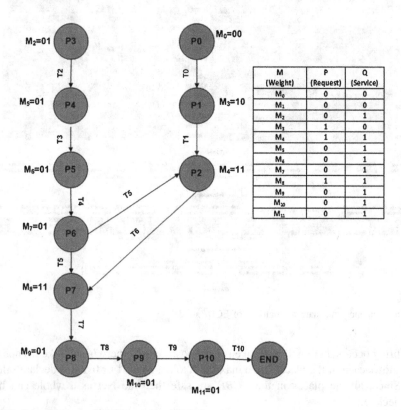

| M (Weight) | P (Request) | Q (Service) |
|---|---|---|
| $M_0$ | 0 | 0 |
| $M_1$ | 0 | 0 |
| $M_2$ | 0 | 1 |
| $M_3$ | 1 | 0 |
| $M_4$ | 1 | 1 |
| $M_5$ | 0 | 1 |
| $M_6$ | 0 | 1 |
| $M_7$ | 0 | 1 |
| $M_8$ | 1 | 1 |
| $M_9$ | 0 | 1 |
| $M_{10}$ | 0 | 1 |
| $M_{11}$ | 0 | 1 |

**Fig. 9** Reachability coverability graph of ECBS

coverability graph starts with initial marking $M0 = [0\ 0]\ T$ and finally reach to state $M11 = [0\ 1]\ T$, where we conclude the service session for the current request of shutdown. As stated earlier, the mark of each place in $P$ of $ECBP$ model will be treated as 1 or 0.

(d) **Liveness**: In the proposed $ECBP$ model as the process starts from $P0$ transitions $T0$ through $T9$ are fired and place $P10$ is reached. The Net is continuous and hence the liveness property is ensured. Therefore the proposed Net is considered to be live.

(e) **Conservativeness**: Conservation property of $ECBP$ model defines that the number of tokens before and after the execution remains constant. Here, sum of all tokens is counted at the initial markings and again in the final markings of each execution. If all the markings in the reachability graph have the same sum of tokens then the Petri Net is declared to be strictly conservative. Hence, $ECBP$ is considered to be strictly conservative.

(f) **Fairness**: Fairness is an important performance criterion for studying behavioral properties of dynamic system. The fairness index always lies between 0 and 1. Here, we have discussed about the bounded fairness or $B$ Fair. An $ECBP$ Net is said to be $B$ Fair Net where every pair of transitions in the Net are in a $B$ fair relation.

# 5 A Case Study

In this work, we establish our approach with the help of a case study of an airline reservation system. Here we provide the Petri Net modeling of $ECBS$ using a tool called PIPE. This application is used to maintain flight details, flight status and reservation process. This Petri Net model can be used by other service industries as well.

Major features provided by the system are:

(a) **Enquiry about the flight details**: The airline reservation system allows the user to perform flight inquiry such as flight scheduling, availability, status, fare details, etc.

(b) **User Registration**: The system allows the user to register as a new user who can book.

(c) **Reservation of Flight**: The system allows the user to book the flights as per their requirements.

Using Tables 4 and 5 we have modeled the roles in the Airline Reservation System that are shown in Fig. 10. The following are the workflow of the proposed case study:

**Table 4** Places and transitions with its descriptions

| Places | Component of places | Transitions | Events |
|---|---|---|---|
| P0 | Customer | T0 | E0 |
| P1 | Booking Portal | T1 | E1 |
| P2 | Airline Server | T2 | E2 |
| P3 | Airline Server | T3 | E3 |
| P4 | Airline Agent | T4 | E4 |
| P5 | Flight Details | T5 | E5 |
| P6 | Airline Reservation Gateway | T6 | E6 |
| P7 | Flight Scheduling Agent | T7 | E6 |
| P8 | Flight Reservation Mapper | T8 | E7 |
| P9 | Flight Reservation Scheduler | T9 | E8 |
| P10 | Airline Reservation Details | T10 | E9 |

**Table 5** Transitions with its descriptions

| Places | Events details | Token | Description of tokens | Token parameters |
|---|---|---|---|---|
| P0 | Customer Request Transmitted | t0 | Request | Sent |
| P1 | Customer Request Provided | t0 | Request | Provided |
| P2 | Flight Details Published | t0 | Request | Registered |
|  |  | t1 | Service | Registered |
| P3 | Flight Details Provided | t1 | Service | Provided |
| P4 | Flight Details Collected | t1 | Service | Published |
| P5 | Flight Details Registered | t1 | Service | Provided |
| P6 | Flight Details Matched | t1 | Service | Collected |
| P7 | Flight Details Discovered | t0 | Request | Matched |
|  |  | t1 | Service | Registered |
| P8 | Flight Details Mapped | t1 | Service | Discovered |
| P9 | Flight Details Scheduled | t1 | Service | Mapped |
| P10 | Flight Details Logged & Dispatched | t1 | Service | Scheduled |

(1) **End Users**:

    (a) Initially, a Customer requests for booking the ticket according to their respective requirements such are source, destination, date, time and the price range.

    (b) These request parameters are passes through the Booking Portal that acts as a provider agent and are registered on Airline Reservation system Gateway.

    (c) Airline Reservation Gateway acts as a *CUDDI* (Cloud Universal Description Discovery and Integration) which is considered to be an end customer.

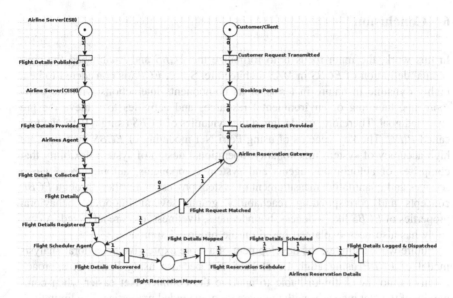

**Fig. 10** Modeling of airline reservation system using ECBP model

(2) **End Service Provider**

 (a) In this architecture an Airline Server which acts as an *ESB* (Enterprise Service Bus) is shown that contains the database of the entire flights that keeps track of the flights and is regularly updated whether new flights are introduced or existing flights are cancelled.

 (b) Another airline server which acts as a *CESB* (Cloud Enterprise Service Bus) is the extension of *ESB* that keeps the record of those flights which are running on the particular day.

 (c) The Airlines Agent which acts like a cloud agent maintains the record of number of seats booked and seats available according to the flight status.

 (d) The entire information is registered under the Flight Details which works as an *HUDDI*. Here, the matching for the required flights is done.

 (e) The Flight Scheduler agent act as a scheduler agent that keeps track the flight schedule.

 (f) Flight Reservation Mapper acts as a Service Mapper which maps the request parameters with the available seats to find the best results.

 (g) Then Flight Reservation Scheduler schedules the required flight as per the customer's choice.

 (h) Finally the Airlines Reservation Details which act as a Service logger will book the best available flight according to the customer's request.

## 6   Conclusion

In this work, the main focus is to formalize the *ECBS* and established the conceptual definition of *ECBS* in *MAS* architecture. Since, *ECBS* is considered to be a highly dynamic in nature in terms of its components interactions with the environments, handling of environmental resources and activities to achieve the pre specified goal. Therefore, we designed the dynamics of *ECBS* using a Petri Net tool called *PIPE*. The key benefits of using Petri Net modeling of *ECBS* dynamics are high accuracy of system functionality, operational design of system flexibility that comprises of autonomous agents and system components, automation of cloud services and quantitative measurement of system processes. Further, through *ECBP* concepts and corresponding reachability graph, several important behavioral properties of *ECBS* like, safeness, boundedness, liveliness, etc. are analyzed. A case study has also been explained based on the proposed approach.

Future work includes, the development of a more expressive behavioral analysis model of *ECBP* using High Level Petri Net because using Petri Net approach tokens cannot be distinguishable to analyze the complex Net easier when complexity of the Net increases with the increase of extended properties. The limitations of the proposed model are inability to support similar kind of service interactions in one single net and less expressive in analyzing the behavioral properties of *ECBS* such as safeness, boundedness and liveliness, etc. in compare to High Level Petri Net.

## References

1. Huhns, M.N., Singh, M.P.: Service-oriented computing: key concepts and principles. IEEE Internet Comput. **9**(1), 75–81 (2005)
2. Arsanjani, A.: Service oriented modeling and architecture. IBM developer works, pp. 1–15 (2004)
3. Zheng, Z., Lyu, M.R.: Collaborative reliability prediction of service-oriented systems. In: Proceedings of the 32nd ACM/IEEE International Conference on Software Engineering, vol. 1, pp. 35–44 (2010)
4. Papazoglou, M.P., Van Den Heuvel, W.J.: Service-oriented computing: state-of-the-art and open research issues. IEEE Comput. (2003)
5. Alexandros, K., Aggelos, G., Vassilis, S., Lampros, K., Magdalinos, P., Antoniou, E., Politopoulou, Z.: A cloud-based farm management system: architecture and implementation. J. Comput. Electron. Agric. **100**, 168–179 (2014)
6. Cavalcante, E.: Architecture Driven Engineering of Cloud-Based Applications. IFIP/Springer-Verlag, Germany, vol. 22, pp. 175–180 (2013)
7. Buyya, R., Ranjan, R., Rodrigo, N.: Intercloud: utility-oriented federation of cloud computing environments for scaling of application services. In: Algorithms and architectures for parallel processing, vol. 6081, pp. 13–31. LNCS. Springer (2010)
8. Brandtzæg, E., Parastoo, M., Mosser, E.: Towards a domain-specific language to deploy applications in the clouds. In: 3rd International Conference on Cloud Computing, GRIDs, and Virtualization. IARIA, pp. 213–218 (2012)

9. Ferry, N., Alessandro, R., Chauvel, F., Morin, B., Solberg, A.: Towards model-driven provisioning, deployment, monitoring, and adaptation of multi-cloud systems. In: 2013 IEEE Sixth International Conference on cloud computing, pp. 887–894 (2013)
10. Divyakant, A., Abbadi, A., Das, S., Elmore, A.J.: Database scalability, elasticity, and autonomy in the cloud. J. Database Syst. Adv. Appl. **2**, 2–15 (2011)
11. Zambonelli, F., Omicini, A.: Challenges and research directions in agent-oriented software engineering. J. Auton. Agents Multi-Agent Syst. **9**, 253–283 (2004)
12. Bauer, B., Mulller, J.P., Odell, J.: Agent UML: a formalism for specifying multiagent software systems. Int. J. Softw. Eng. Knowl. Eng. **11**(3), 1–24 (2001)
13. Klügl, F.: Measuring Complexity of multi-agent simulations—an attempt using metrics. In: Languages, Methodologies and Development Tools for Multi-Agent Systems. Springer, Berlin, Heidelberg (2008)
14. Dhavachelvan, P., Saravanan, S., Satheskumar, K.: Validation of complexity metrics of agent-based systems using Weyuker's axioms. In: International Conference on Information Technology (ICIT '08), pp. 248– 251 (2008)
15. Sarkar, A., Debnath, N.C.: Measuring complexity of multi-agent system architecture. In: 10th IEEE Conference on Industrial Informatics, pp. 998–1003 (2012)
16. Sarkar, A.: Modeling multi-agent system dynamics: graph semantic based approach. In: 10th International Conference on Service Systems and Service Management, pp. 664–669 (2013)
17. Djamel, B.: An agent-based approach for hybrid multi-cloud applications. J. Scalable Comput. Pract. Experience **2**(14), 95–109 (2013)
18. Mandal, A., Changder, S., Sarkar, A., Debnath, N.: A novel and flexible cloud architecture for data-centric applications. In: IEEE International Conference on Industrial Technology (ICIT), pp. 1834–1839 (2013)
19. Khan, G., Sengupta, S., Sarkar, A., Debnath, N.C.: Modeling of inter-cloud architecture using UML 2.0: multi-agent abstraction based approach. In: 23rd International Conference on Software Engineering and Data Engineering, pp. 149–154 (2014)
20. Khan, G., Sengupta, S., Sarkar, A.: Modelling of services and their collaboration in Enterprise Cloud Bus (ECB) using UML 2.0, 2015. In: International Conference on Advances in Computer Engineering and Applications, pp. 207–213, India (2015)
21. Khan, G., Sengupta, S., Sarkar, A.: WSRM: a relational model for web service discovery in Enterprise Cloud Bus (ECB). In: 3rd International Conference on Eco-friendly Computing and Communication System, pp. 117–122, India (2014)
22. Khan, G., Sengupta, S., Sarkar, A., Debnath, N.C.: Web service discovery in enterprise cloud bus framework: T vector based model. In: 13th IEEE International Conference on Industrial Informatics, pp. 1672–1677 (2015)
23. Khan, G., Sengupta, S., Sarkar, A., Debnath, N.C.: XML based service registration system for enterprise cloud bus. In: 3rd IEEE International Conference on Computing, Management and Communications, pp. 250–255 (2015)
24. Khan, G., Sengupta, S., Sarkar, A.: Priority based service scheduling in enterprise cloud bus architecture. In: IEEE International Conference on Foundations and Frontiers in Computer, Communication and Electrical Engineering (C2E2 2016), SKFGI, pp. 363–368, Mankundu, India (2016)
25. Khan, G., Sengupta, S., Sarkar, A., Debnath, N.C.: Performance analysis of service discovery for Enterprise Cloud Bus (ECB). In: IEEE International Conference on Industrial Technology (ICIT 2016), pp. 1728–1735, Taipei, Taiwan (2016)
26. Bloom, J., Clark, C., Clifford, C., Duncan, A., Khan, H., Papantoniou, M.: Platform Independent Petri-Net Editor: Final Report, London (2003)
27. BoNet, P., Llado, M.C., Puigjaner, R.: PIPE v2.5: a Petri Net tool for performance modeling. In: Proceedings of the 23rd Latin American Conference on Informatics (CLEI 2007), San Jose, Costa Rica (2007)
28. Marzougui, B., Hassine, K., Barkaoui, K.: A new formalism for modeling a multi-agent systems: agent Petri Nets. J. Softw. Eng. Appl. **3**(12), 1118–1124 (2010)

29. Celaya, J.R., Desrochers, A.A., Graves, R.J.: Modeling and analysis of multi-agent systems using Petri Nets. J. Comput. Academy Press **4**(10), 981–996 (2009)
30. Chainbi, W.: Multi-agent systems: a Petri Net with objects based approach. In: IEEE/WIC/ACM International Conference on Intelligent Agent Technology (2004)
31. Pujari, S., Mukhopadhyay, S.: Petri Net: a tool for modeling and analyze multi-agent oriented systems. Int. J. Intell. Syst Appl. **10**, 103–112 (2012)
32. Chatterjee, A.K., Sarkar, A., Bhattacharya, S.: Modeling and analysis of agent oriented system: Petri-net based approach. In: 11th International Conference on Software Engineering Research and Practice (SERP 11), vol. 1, pp. 17–23 (2011)
33. Sofiane, B., Bendoukha, H., Moldt, H.: ICNETS: towards designing inter-cloud workflow management systems by Petri Nets. In: Enterprise and Organizational Modeling and Simulation, vol. 198, pp. 187–198. Springer International Publishing (2015)

# Service Provisioning in Cloud: A Systematic Survey

**Adrija Bhattacharya and Sankhayan Choudhury**

**Abstract** Cloud Computing is all about delivering services over the Internet. It has some technical, business and economical aspects. The complexity of service provisioning has increased significantly with the increasing number of cloud services and their providers. This creates a complex situation and as a result the service provisioning techniques face hurdles. The challenge of service provisioning is to properly offer services by adjusting the complexities efficiently. There are significant works to solve the problem in different manners. But still there are some gaps that are to be noticed and bridged for future advancement of cloud technology research. In this paper an attempt has been made for analyzing the service provisioning techniques from different perspectives. The said perspectives are various techniques and methodologies, QoS parameter considered, context awareness, etc. Moreover, the role of a broker in this context is also addressed. The overall motivation is to identify the open challenges, that may provide a future research direction in context of service provisioning.

**Keywords** Survey · Service provisioning · Broker · Cloud

## 1 Introduction

The growing nature of cloud industry has raised the need to discover newer techniques to efficiently handle cloud providers and services. According to NIST Cloud reference architecture in [1], Cloud computing deliverables must follow any of the three service models namely Infrastructure as a Service (IaaS), Platform as a Service (Paas) and Software as a Service (SaaS). In case of IaaS service providers responsibility is limited to manage different processing, physical infrastructures like storage

A. Bhattacharya (✉) · S. Choudhury
Department of Computer Science & Engineering, University of Calcutta, Kolkata, India
e-mail: adrija.bhattacharya@gmail.com

S. Choudhury
e-mail: sankhayan@gmail.com

© Springer Nature Singapore Pte Ltd. 2017
R. Chaki et al. (eds.), *Advanced Computing and Systems for Security*,
Advances in Intelligent Systems and Computing 568,
DOI 10.1007/978-981-10-3391-9_3

or network etc. for a particular set of consumers. PaaS provides development and monitoring tools, appropriate middle-ware etc. SaaS is sole responsible for handling software needs and related management tasks for consumers. This large number of heterogeneous (with respect to tasks and technicality) services are often efficiently handled by different intelligent techniques. The service provisioning starts with discovery of relevant services according to user requirements. But the complexity of the problem increases with the increase in number of service providers. Often there exist multiple services with the same functionality. It opens a problem of service selection i.e. comparing the similar services with respect to their quality aspects. As the service selection is based on huge service data; as a result the complexity of the problem increased to exponential one. Service provisioning addressed another well known problem of service composition. Service composition is the technique to combine two or more service together as a workflow towards the fulfillment of user requirement. It is a technique to provide more consolidated service sets maintaining compliance, QoS, etc. In general the problem of service composition is considered as an optimization problem. It tries to optimize set of QoS parameters as per consumer's requirement and also to select appropriate and optimal simple services to consolidate them into composite complex services. The varying choices of QoSs from different customers makes the service composition problem a critical one. Moreover, the enormity of the solution space make the multi object multi constraint optimization problem as a NP-hard one.

Service provisioning solution in general is introduced through a concept of broker. The idea of brokerage was far introduced in the domain of web services [2], but it is incorporated into the area of cloud computing. Brokers are basically third party entities between provider and consumers to make the coordination easier. It is some way similar to Multi-source Integrator (MSI), but there exists differences too. Cloud service brokers (CSB) are more efficient than MSI in handling huge number of terms and conditions with respect to service provisioning. CSB has more usage in a truly distributed environment that was not at all achieved by MSI. Brokers can work in all of the cloud architecture like private, public or hybrid. There exist different customized versions of brokers designed for various application domains. It is often described as a single interface provider through which multiple clouds can be accessed and resources may be shared [3]. It sometimes includes the cloud failure detection mechanisms and works outside the clouds. The benefits have been realized during a continuous evolving process of improvement in brokers role and it is a continuing till date. A broker has information about business processes, existing service from service providers, requirements of a consumer in detail. It operates accordingly on the information and tries to find out the best possible solution suitable for the requirement. Besides, it works on other sides like reporting, billing, auditing etc. and provides a complete integrated solution at one point [4]. That particular solution is looking like Fig. 1.

In general the brokers have duties like, service provisioning, cloud security, SLA negotiation, reporting, billing etc. But it helps in reducing the cloud complexity and the burden on consumers. The failure assessment and change management is another two important goals that broker often carries. The performance assessment

**Fig. 1** Outline of brokerage

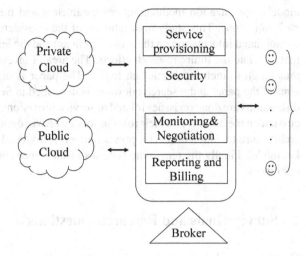

of an integrated solution is together with respect to heterogeneity involved within the multi-cloud environment. Similarly the decomposition of reasons behind the performance degradation is a crucial issue in multi-cloud environment. Handling the frequently changing business processes involved any service composition or discovery process is another challenge that is partially managed by the brokers. It may seem a little confusing to decide the responsibilities of a broker within a single umbrella as it has many roles in different domain with different level of abstraction. Gartner [5] summarized the different roles of cloud broker in three broader categories. These are Aggregation, customization and integration. Whatever may be the role cloud broker play, it belongs to any of the three categories. According to service management perspective broker has three types of responsibilities that constitute of business support, configuration and interoperability issues. Broker handles the aforesaid issues in a manner that none of user and providers gets information about this. i.e. broker basically creates an abstraction between service producer and consumer.

The architecture of a broker is not at all standardized. The service provisioning role of broker demand serious attention and modification in design. Here in this paper we have tried to comment on that and also tried to identify which components are essential for service provisioning. We confined our discussion to the service provisioning functionality of a broker and related issues handles by researchers and some open challenges that are identified in recent years despite of several other activities by broker. This survey provides a systematic literature review on state-of the art approaches and techniques used in Cloud Service provisioning by broker. The advancements so far on this topic are discussed to provide useful information to motivate further researches in this area. Identification of different objectives along with reasonable taxonomy definition of divergent approaches and mechanisms is a major achievement of the review. A set of classifications are done following different approaches on ranking methods, prediction methods, performance assessment models, optimization methods, query parameters and metrics on which the comparison is done. This study highlights all of the considered QoS parameters, gives consolidated information on the most significant and the least considered parameters that alleviate the future research efforts. The area of investigation is very extensive, thus tough to include all relevant topics. The paper is organized as follows. Main aims of the paper and research questions is described in Sect. 2. In Sect. 3, extensive discussions are done on issues related to service provisioning. Section 4 contains discussion on the brokers and their roles in service provisioning. In this section a general architecture for efficient service previsioning is introduced along with proposed broker model. Finally the remaining sections of the paper contain the conclusions and references.

## 2 Survey Goals and Research Questions

Current research have tried to address the significant issues of service provisioning and role of broker in this context. More specifically, the extraction of underlying features and methods used in many approaches are being considered. The study has approached extensively for identifying representative works from service provisioning; namely service discovery, Selection and composition strategies. Most researchers have considered QoS parameters and proposed objective functions in terms of those QoSs. In this course of action the following questions are addressed in this paper

Q 1. Whether the purpose of research is fully defined in the papers or not?
Q 2. What is the proposed approach and what are the methods used defined clearly?
Q 3. How have the researchers conducted the research?
Q 4. What procedures discussed are supported by the assessment results in each paper?
Q 5. What QoS parameters are accounted for?
Q 6. Awareness about the properties of the technology used?
Q 7. Explained the details of underlying Architecture?

Around 80 papers that were studied from 2009 to December 2014 from different high-level refereed journals and prestigious international conferences. After studying and comparing we have pointed out some future research ares that are summarized in Conclusion Section.

## 3 Detailed Discussion on Service Provisioning

Here in this paper, the service provisioning role of a cloud service broker is under consideration. The different aspects of classifying the existing works on service provisioning by broker is studied and taxonomy is identified among these works. The following subsections have different angles of the survey on architectural specification

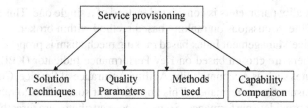

Fig. 2 Aspects of the survey

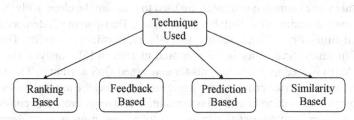

Fig. 3 Selection techniques used

for service oriented computing in cloud domain, different mathematical techniques used for modeling, performance measuring parameters used, etc. Another way of efficient service provisioning can be handled by cloud service brokers. The Fig. 2 outlines the aspects considered for the classification and specifying the identified taxonomy. Broker based provisioning approach is discussed in detail in Sect. 4.

Now the survey is done from the above four aspects. There exists more sub categories under each categories in Fig. 3. Following sections and Subsections discuss the literature reviews in detail.

## 3.1 Solution Techniques Used

Several mathematical techniques are used on cloud services selection within broker framework for different measures. There is another taxonomy identified among the methodology of mathematics used. There is a figure (Fig. 3) that summarizes different methods used in this subsection.

In the figure there are four types of selection methods used. In following subsections each of the method is discussed in detail.

### 3.1.1 Ranking Based Methods

Services are often ranked based on different parameters with different purposes. There are some of the approaches that ranked to decide the most suitable service provider based ont heir quality performance. Other approaches ranked services with respect to cost, often with respect to reliability etc. Sometimes a combination of some of the quality parameters is considered instead of a single one. This subsection constitutes of the discussions on ranking based method within broker.

In [6] Service Management Index based ranking mechanism is proposed. Here the service providers are chosen based on Key Performance Indicator (KPI). Different quality attributes such as Accountability, Agility, Assurance of Service, Cost are considered to measure the performance. This approach put some effort to rank services based on their providers performance. Some of these attributes are directly related to the performance of customer application and others are related to provider's performance. Only some quantifiable metrics are used to measure the cloud KPIs. Service response time, Sustainability, Suitability, Accuracy, Transparency, interoperability, Cost, Availability, Reliability, Stability, Adaptability, Elasticity, Usability, Throughput and Efficiency, Scalability is the proposed metrics in [6]. Analytic Hierarchy Process (AHP) is used to define weights for concerned QoS attributes. The relative ranking values for service providers are aggregated to select the top ranked services.

The objective of ranking here in [7] is to maintain cost requirements of consumers. The best provider with respect to cost can be found out in many ways depending on other related quality parameters such as bandwidth, storage, etc. This paper also includes the newer idea of deciding the ranking based on the quality parameter values by comparing with the benchmarks. The actual deviation from that standardization of quality parameter is the key factor behind service ranking.

In [8] a ranking methodology proposed based on Cloud Service Provider (CSP) index. Here properties that affect the values of providers CSP index are identified. A B+ tree based approach is proposed for speeding up the query processing. Different quality properties are encoded and fair enough relationship also exists among the property encodings. K-mean algorithm is used for clustering services according to their close offerings of quality parameters. Service are found based on the user query and relevant services are ranked according to the Hamming distances from the exact point of need within the cluster.

In [9] an approach is discussed that decides the efficiency of services based on the affinity between the Software and platform as a Service Provider (SP) and Infrastructure Provider (IP). Here SP has a preference list of IPs considering the issues such as Past Performance, Maintenance, Infrastructure, Security and Customer Support of IPs. On the other hand IPs are also have preference list based on Past Performance, Legal and Security.

Ranking services according to performance is a common key criteria to choose among similar services. Reference [6] considers relative ranking among services providers. The KPI based [6] approach is suitable for comparing among the service providers than CSP index [8] based approach. In terms of query processing time [8] would perform better. The SP and IP affinity based ranking is an efficient one but it invites the vendor lock in situation in cloud. The kind of inter-dependencies among SP-IP would compel customer to choose unsuitable (high priced and overloaded) services. Besides the above approaches, parallel research works exist where other metrics and methodologies are used. For example in [10] a usage pattern based ranking approach is described. This is an area where more attention can be given on real deployment of services and related predictive usage needs from consumers.

### 3.1.2  Feedback Based

In [11] a framework has been designed that helps in cross cloud migration of software service. The cloud service brokers role for handling the interoperability issues. After deploying the framework in actual working environment, the impact of this has to be recorded and considered for further rectifications. The feedbacks are immense helpful for developers to handle the maintenance issues, usability etc. for the frameworks. This improves the testing process of the framework. In [12] an attempt has been made to select best services with respect to users query based on QoS parameters. Unlike other methods, it uses user related information to obtain the QoS of a service. It relies on both customer feedback and system generated feedbacks so it is more accurate than [11]. The QoS of already consumed services are estimate and the service selection mechanism is executed and tested by objective performance benchmark testing. It is the first approach towards the advancement and testing of feedback based service selection. In [13] a feedback based method for weather forecasting services is proposed. Here the full customer requirements and priorities are set based on their feedbacks. This is useful in any service oriented context as the QoS requirements often contain conflicting QoS parameters. The uncertainties regarding the changing prior objectives of users is dissolved by the feedback mechanism. Reference [14] illustrates a framework where users give the feedbacks as Quality of Experience. The functional and non-functional requirements demanded by users were at all actually met by the provided services or not; that is the key issue here. User poses their past experiences based on opinion and preferences about the specification of services as one of the inputs to the system in the next round. This helps to efficiently match further user requirements with the performance of the services as needed. Another similar kind of work has been done in [15], where user maintains a history of evaluations of the provided service quality. This history can be used in future for a better service selection process. Among all the approached [12, 14] are two approaches that are closer to grab consumer's feedback not only with respect to QoSs but the satisfaction of users also.

### 3.1.3  Prediction Based

Reference [16] illustrates a general framework for cloud service provisioning. QoS satisfaction of offered are predicted through an HMM based model. The components considered within a composition plan collectively satisfy the user required QoS or not is the main objective of the work. There also exists provision for considering user feedback based on QoS experiences to match future composition of services. Often it is needed in real time applications to meet hard deadlines. These deadlines are expressed in terms of QoS of the services. Here a prediction model in [13] is used to calculate possible costs and task decomposition for meeting user deadlines. A feedback based method is used for determining the best suitable service based on user defined deadlines that is different cost, execution time, etc. In this paper another described method for doing the same is by setting priorities to different constraints and these are satisfied by different services. A layered approach is described here and the interoperability among the layers and the prediction models enable the scope for provisioning services according to users high level requirements. The resource provisioning services in cloud computing domain demands a dynamic formalism that can adapt with the requirement of user. The adaptability will work based on the knowledge from history that is being updated time to time [17]. Multiple execution plans exist there and the service selection is based on their predicted cost. The correlation between the resources and recent operation (like consumption etc.) history is considered within a prediction model in [18]. The correlation calculated based on previous data. In case of newly developed operation on resources the prediction is based on a linear regression model. In this context, another history based service profiling method in [19] is there that helps in predicting future method invocation based on past invocation data. Here the cost analysis also done by mapping network bandwidth and latency into cost. The cost prediction here plays an important role in service profiling and selection. The methodology used in [20] is a prediction model for forecasting the behavior of conflicting goal parameters (energy consumption, quality, etc.). Already we have seen a ranking methodology based on conflicting parameters in [13]; but here the service selection decision is taken here based on the prediction. The prediction here is tough due to the reciprocal relationship among chosen parameter; but the model designed works to balance among the parameter and predict accordingly. Prediction efficiency is the key criteria for this approaches. The more the approaches are dedicated to history bases efficiency increases. Some of the approaches as [13, 18] are only interested in cost prediction. But the other QoS s are also important to predict for more useful selection of services. In this regard [17, 19] are advanced works in this domain.

### 3.1.4 Similarity Based

Similarity checking in service oriented computing can be seen in many aspect. Such as, similarity among functionality of services, QoS demand similarity, Similarity in usage pattern, similarity based performance applications, etc. This subsection discusses some of the similarity based mechanism that already been used within cloud domain or can be employed within brokers. In [21] brokerage approach is described based on a benchmark similarity classification method for measuring application performance. The similarity measure among the QoS parameters and input and outputs of different services is done in [22]. These similarity quotients are stored. Based on that measure the service composition graph is formed and similarity quotients work as weights there. The different quality claims as well as users demand specification plays an important role in service provisioning. Similarity matching mechanism for classifying service demands and an algorithm is designed for user requirement normalization in [23]. An algorithm is developed in [24], that generates a composition graph of service and then a similarity measures of services is done. The composition system is finally built based on similarity of services for identifying service interfaces for provisioning. In [25] for best suited service selection, initially a complete list of services a similarity based approach is introduced. After that similarity computation steps preference list of QoSs is prepared. Those QoSs are predicted and based on that ranking is done. Similarity among user query and service advertisement is done in [26]. Here the key methodology for similarity matching is based on OWL-S engine. The broker is the sole responder for finding query specific services and acquiring providers specifications.

## 3.2 QoS Parameter Used

Some or all of the research works discussed in Sects. 3.1 and 3.2 have Quality of Service (QoS) parameters mentioned. Some of these parameters are Cost, Performance, Reliability, Security, Bandwidth and Latency (Especially for Network services), etc. Now we are listing out some of the other researches on quality awareness in service oriented cloud domain that are either already included or can be included within broker framework. Table 1 summarizes the QoS attributes, reference number and

**Table 1**  References with publication year and mentioned QoS parameters

| Reference | Year | QoS parameters |
|-----------|------|----------------|
| [27] | 2012 | Throughput, response time |
| [28] | 2012 | Latency, cost, reliability, availability |
| [9] | 2012 | Efficiency, security, trust |
| [29] | 2010 | Availability, cost, reliability, execution delay |
| [30] | 2011 | Reliability, availability, cost |
| [31] | 2013 | Security, cost, availability, energy consumption |
| [32] | 2011 | Cost efficiency, security, trust |
| [33] | 2014 | Cost, availability |
| [34] | 2012 | Reliability, deployment, network related QoS, latency |
| [35] | 2013 | Time, cost, availability, reliability |
| [36] | 2013 | Cost |
| [37] | 2013 | Performance, reliability, cost, rating, integrity, security |
| [38] | 2013 | Response time, cost |
| [39] | 2013 | Response time, cost, availability, reliability |
| [40] | 2013 | Cost, latency, reputation |
| [41] | 2013 | Cost, response time, security, reputation, availability, reliability, Durability, Data Control |
| [42] | 2011 | Response time, cost, throughput, reputation, availability, Reliability |
| [43] | 2011 | Response time, cost, availability, reputation |
| [44] | 2013 | Response time, throughput |
| [45] | 2013 | Response time, cost, reliability, efficiency, trust, maintainability |

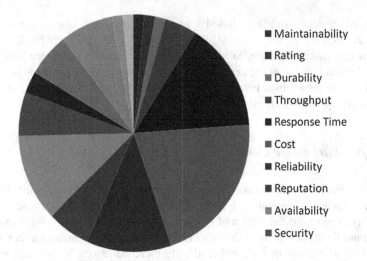

**Fig. 4** Proportion of QoS parameters

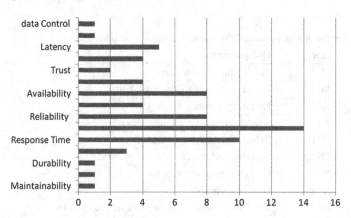

**Fig. 5** Importance of QoS parameters

publication year. Fig. 4 signifies the proportions of QoS covered by the references in Table 1. Fig. 5 shows the relative importance of those QoS parameters.

A few recent works that emphasizes on QoS parameters are listed in the Table 1. The entries signify the popularities of QoS parameters in the following diagram Fig. 6.

Figure 7 has the reflected the importance of a QoS according to our study. Though the cost parameter may have several measures other type of cost such as, data transmission cost, Virtual machine Migration cost, Appliance cost, Energy cost, etc. still in overall consideration the Cost parameter is the most important one according to our study.

Besides the above mentioned QoS, there exists a few more. Such as robustness, scalability, performance, transactional, obligation, etc. In the future works for assess-

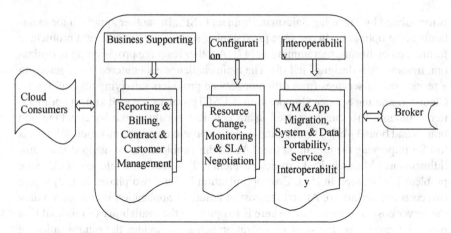

**Fig. 6** Broker's responsibilities

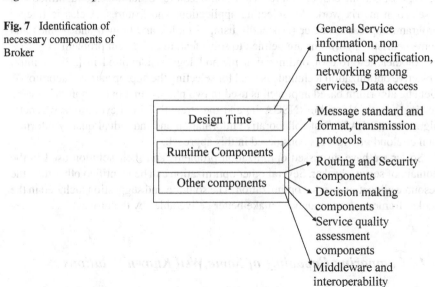

**Fig. 7** Identification of
necessary components of
Broker

ing QoS and service selection strategies in cloud will consider some of these newer
parameters.

## 3.3 Methodologies Used

Service oriented capabilities has analytical power for not only handling data but also
used for handling large optimization and decision problems. In [46] service provisioning in multi-cloud environment the service acquiring process selection problem

is formulated by Traveling Salesman Problem (TSP). In another approach for maintaining cost optimality in resource provisioning service the optimization problem is formulated by linear programming in [47]. Another resource provisioning optimization framework is designed in [48]. The main challenge encountered while selecting a service out of services from different service providers satisfying all QoS levels. Cloud Computing Service Composition (CCSC) problem is a well known optimization problem. This can be solved using classic algorithms like backtracking and branch and bound [49]. The mapping of such a problem and optimization solution is just for improving the execution time. Again this problem can be mapped into Multidimensional Multi choice knapsack problem [50]. A modified solution of the same problem is discussed in [51]. Cost optimization is done two phases in [52] where two costs are optimized. At first the cost of failure of application is considered then the server cost. The main motive here is to optimize the availability of minimal cost non-strict resources. Two-way optimization actually provides the minimization of server cost and the cost of failures. Here some features of failures are identified and a selection matrix works for selecting applications and features. A classic Integer programming based solver (especially using Simplex and branch and bound) and some heuristic algorithms are defined to solve feature placement problem.

Another new role of optimization methodology is identified in [53]. A multi objective optimization technique is used for selecting the appropriate collaborator. A Genetic algorithm based approach is used in two phases; first one is non-dominated sorting genetic algorithm (NSGA-II) and the strength Pareto evolutionary genetic algorithm (SPEA2). Past collaborative information and individual quality information of cloud providers is considered in this approach.

So far we have discussed optimization problems and their solution used in the domain of service broker. Several other optimization can be identified other than the resource allocation service optimization. The solution and suggestion included in the broker framework cane anyway make better deliverables by the brokers.

## 3.4 Comparing Capability of Some Well Known Solutions

There exists a few approaches of brokerage that provides sole solutions to the existing problems in service provisioning by brokerage. First of all there exist a RESERVOIR [21] Framework is a decentralized federation. It has some collaborating business organizations. The architecture does not have any direct communication among each entity of the framework. It is the most popular one. We have done a comparative study among RESERVOIR [21], CloudFoundry [54], CompatibleOne [4] and OPTIMIS [9]. Table 2 compares these in a systematic manner.

There exists some of the performance monitoring approaches that are very much popular now a days. We have done a comparative study among those popular monitoring services and the following table reflects the outcome. Table 3 summarizes that which monitoring services has considered which properties.

**Table 2** Well known solution comparison

| Reference | Scalability | SLA monitoring | Programming environment | Interoperability |
|---|---|---|---|---|
| [21] | ECA rules to scale infrastructure | Amalgamates IP multicast along with (JMS) publish-subscribe system | JMS & WASUM monitoring Platform | YES |
| [9] | Includes Elasticity Engine | Uses REST to get CPU/disk usage from monitoring | Java schemas jaxb, xmlbeans, REST, monitor; also jaxws, cxf, javagat. | Through OpenNebula & OpenStack Manager |
| [4] | Only load balancer module for the IaaS resources | COMONS Monitoring module | PaaS4Dev: Java EE (EE5/6 web profile) & Enterprise OSGi services | PROCCIs OpenNebula OpenStack & Azure |
| [54] | Can add/remove instances for scalability & increase/decrease CPU | Basic logging facility with Cloud third party monitoring | RUBU & GO | Supports AWS, vSphere, OpenStack & Rackspace |

**Table 3** References and their monitoring parameters

| Reference | Monitoring parameters |
|---|---|
| [55] | Timeliness, reliability |
| [56] | Timeliness, comprehensiveness |
| [57] | Availability, accuracy |
| [58] | Availability, accuracy |
| [59–62] | Availability, timeliness, resilience |

**Table 4** References with supported cloud characteristics

| Reference | Supported cloud characteristic |
|---|---|
| [63] | Extensibility |
| [64] | Scalability, adaptability |
| [65] | Timeliness |
| [66] | Autonomicity |
| [67, 68] | Adaptability, extensibility |
| [69] | Scalability |
| [70] | Elasticity |

In the next option for comparison we have considered the different properties of Cloud computing that are supported by several popular implementation platforms in Table 4.

In this subsection we have seen some comparative study among well known solutions. Here in Table 3 we can see that the monitoring parameters are not wide variant. The parameters are very much confined to Timeliness, Accuracy and Availability. But there are other very important properties like security, Response Time that has to be considered for monitoring to judge the performance of cloud services. In Table 4, it is reflected that the Scalability and Elasticity could not get more attention while implementation. Though these two parameters are the pillars of service provisioning in cloud. Along with this two Adaptability must also get more attention to provide actual flavor of cloud federation which is the next generation technology.

## 4  Context Awareness in Service Provisioning

Context is information that can describe the situation of anything [71]. It is often defined in terms of [72] answers to five "wh" questions (Who, What, Where, When, Why). Context adds more intense meaning to already existing information about any surrounds [73]. The new generations of all sensor based system are highly dependent on the context awareness property of computing and data retrieval mechanisms; due to the reason that the sensor data interpretation changes with the associated context scenario. It is particularly important to grab the context such that it would help to identify the right situation and dependent actions can be taken. The services relevant for a particular context may seem completely irrelevant in a slightly different context. So the customization of service provisioning depending on the context situation is the next open area of research.

A large number of cloud based system in Pervasive environment puts various challenges containing heterogeneous applications and Software systems. The sensor based mobile and cloud environments have some crucial open issues that have to be addressed with respect to the context aware service provisioning. Sensing as a service is developed to satisfy huge set of service consumer requirements. Cloud computing has a flexible design and huge computing resources. The amalgamation of Cloud Technology along with the Sensing capabilities are pioneering the area of context aware software systems. It also enhances the overall performance by adding mobile devices (convenient ubiquitous interface), and a backbone infrastructure with huge computing resources. It has a set of newer challenges compared to older service provisioning motivations. The concerned area of broker based service provisioning opens a newer area that includes context awareness in cloud based environment. The significant works in the area are classified with respect to the following aspects

1. **Context Management** Context management refers to identifying the process of context data collection to production of that data as a useful piece of information. The discussion of context management is incomplete without referring some of

the work that provides context information as a service. The first part of the management is to collect data from appropriate sensors. The context information is either of two types; namely discrete or continuous. In [74] the context data are gathered based on some related events. The type of event (discrete or continuous) determines the type of context acquisition. The events like Door open or lights on are discrete events; the data acquisition from sensors depending on this events will be discrete in nature. The event such as Raining or driving needs continuous data acquisition from sensors. Most of the significant works as [75–78] outlined context management in three or four steps. The steps are context data acquisition, processing and representation and decision making. Alternately some of the approaches [79, 80] have considered many steps towards the final representation of context information. But to standardize the process of context data management it ideally can have four stages namely collection, Processing, Analysis and Decision.

In [81] a framework for context information exchanges between service consumer and provider is proposed. Context brokers are there to mediate. Relevant services are selected based on a multi criteria decision algorithm. In this approach a QoS is used to derive the quality of the context information. Moreover,the proposed algorithm calculates the scores for services according to the requirements of consumers. In [82] a framework for maintenance of interoperability and independence is discussed. This approach has mechanism to processing heterogeneous context data and produce meaningful high level context information. ContextML and Context Query Language both are proposed in this paper that makes service provisioning algorithm efficient according to required consumers context.

2. **Architecture Supporting Context Awareness** Supporting infrastructure for generating context information is very important. Automatic and adaptive application development requires stronger architectural base to support efficient context data acquisition, processing power for producing relevant context information and meaningful high level representation that may facilitate other application's requirement of context awareness. In [83] CASS (context aware simulation system) is proposed. It is a self-adaptive application developer for producing context information checking conflicts and achieve optimality in using data acquisition sensors. A similar approach is discussed in [84]. This approach deals with the availability of sensor data which is very crucial in any smart monitoring application. Another type of architectural components contributed by many researchers is a context aware middle ware. Two approaches [74, 83] have significant contribution. In [74] some hierarchies among the context information are defined. This helps in gradual improvement of context information. It is based on object modeling; same as [83]. But this approach is more generic than that in [83]. But the need of typical adhoc sensor network is more fulfilled by [83]. It also supports security and produces context result based on relational data model which is more accurate compared to ontology based approach of [74]. Reference [85] is a very popular agent based architecture comprising of context brokers for determining context in intelligent spaces (sensor monitored locations). It works upon OWL-based ontology and aggregates the context information successfully as needed.

In the due course the middle ware systems used must be equipped with technologies for sharing context among heterogeneous applications resulting increase in re usability. User experience, feedback, preference, must be included in the newer algorithms of service discovery and selection, ranking as synergy with context information. Classic keyword based discovery approach fails to grab the dynamicity of service requirements and it is ineffective in retrieving proper service information. Collection, modeling, reasoning, distribution is necessary stages of service context life cycle. Standardization of this context aware architecture is needed to efficiently provide services to the consumers. The RESTfulness in services extends the area of service provisioning confined to web only towards any pervasive environments. Security and privacy are also important area to contribute in context aware systems as there exists huge number of sensors in any pervasive environment.

3. **Context aware Service Provisioning** In an intermediate stage of research regarding the service discovery techniques; some location aware mechanisms were devised [86, 87]. These used the geo-spatial information for sensing proximity. In context aware discovery, proximity extends beyond just the geographical location [88–90]. In [103] researchers proposed the EASY (Efficient semantic Service discovery) framework which considers QoS and context for service discovery. It uses on OWL-S for context description. But this framework did not consider the importance of continuous context acquisition. Thus the efficiency of context aware discovery is questionable. In [104] proposed a mechanism that uses discovery scope and user dependent context information to grab the change in service need and according provisioning. Similar approach in [91] and [105] gives a user specific context acquisition and the choices of service changes with the change in context. In [92] a comprehensive formal model of context is proposed. Here the service discovery algorithm copes up with the changes in context and preferences and provision services accordingly in real time. The context aware service provisioning approach must be either light weight or must be capable of offloading computing part on top of resource-rich cloud computing infrastructure. A hybrid approach has to be the most appropriate solution for this. Context based performance analysis and ranking of services needs attention of researchers.

## 5   Service Provisioning by Brokers

The discussion on service provisioning is incomplete without discussing service brokers. Almost all of the aspects of better service provisioning described so far, can be deployed well by the broker based design. Gartner classified all of the broker roles for service provisioning within three categories. This classification assumes only the role of a broker as a service composer. The following subsection detailed out the categories namely Aggregation, Integration and Customization. In case of aggregating the services broker actually composes two or more service from one or more providers and then delivers it to the user or any other service provider. This type of brokerage helps in centralized service management, normalized service discovery and access, SLA management [93] etc. But this don't add any new functionality though, it provides a well managed abstraction for both service providers and consumers. Marketplace and Appstore are the real life example of cloud brokerage with respect to aggregation. Integration by service broker can be of any type like cloud-cloud integration, supply chain integration etc. This integration often leads to substantially new values to community management in cloud domain [94]. It allows process integration that yields independent services that are mostly implemented as PaaS. Real life example of this kind is contacts of gmail. In case of customization, newer functionalities are added to improve the existing service functionality. It modifies the original cloud services and implements newer application [95]. This happens often in developing a newer composite application in process and data enhancement. But there exists different other roles. In another aspect the business oriented thinking produces the classification based on service management associated with service provisioning. Figure 2 summarizes that from this point of view there are another three categories that are Business support, configuration and interoperability. Section 3.4 discusses this in details.

## 5.1  Broker's Responsibilities

This section illustrates the responsibilities of brokers in detail.

### 5.1.1  Business Support

The service consumer in a cloud environment may be another cloud service provider itself or an end user. There exist a few responsibilities from broker side to handle customers in an efficient way. The responsibility of broker starts from accepting a query. Then it works as a contract builder, search manager, calculating costs involved and final billing, pricing and service ratings are the customer related duties of a broker.

### 5.1.2  Configuration Management

Main duty in this layer is the SLA negotiation. After that the phase of actual service composition starts. In this category broker takes responsibility of resource provisioning and construction of required business plan takes place here. Performance related measurement and quality estimation is an important one to mention. Error monitoring is another customary of the broker in this role. After monitoring errors proper reporting and resolving them are within the scope of this category.

### 5.1.3 Interoperability

This is the most important part. As in this layer the actual service offering starts. Initially checking with the compatibility issues service broker moves towards the data transfer process. By this, actual service information and the service functionalities are accessed by consumer. The ultimate step a broker monitors and executes is the migration phase. That may be VM migration in case of IaaS and PaaS and Apps migration in case of SaaS.

## 5.2 Service Broker: Architectures

Broker is a very essential component in service oriented architecture. In cloud computing; particularly in the multi cloud environment typically broker is needed to ensure service provisioning. The standard cloud service broker architecture is yet to find; but there exists a set of different service broker working in the cloud domain with varying architectural properties. In most cases, the brokers are designed to fulfil a purpose specific to the application domain. In this subsection, we will discuss some of those architecture and their pros and cons. Brokers work for service provisioning, SLA negotiation, Reporting, billing, etc. but here we are interested in the service provisioning role of a broker. Again, generally service broker has some specified roles. These roles can be broadly classified in management issues, security issues, business issues and other issues. It is clear after some literature reviews that all these issues are not mandatory for every broker to handle. It is highly dependent on the application domain that where it is to deploy and what are the essential roles a broker must play at a particular situation. Thus the architecture varies from one broker to another depending on the responsibilities it carries.

The heterogeneity and diverse requirements in network service applications are the main reason behind discovering a newer approach. A layer between services and user application is developed in [96] that named network service provisioning layer. It has two parts as network service broker and service registry. Moreover it has components that have capability of deciding achievable connectivity and QoS of the network services. Approach in [96] has a detailed description of a structure for network service description and discovery. This work has the facility that it provides partial description publishing option that improves network description, discovery and resource allocation. Moreover, the implementation is independent of platform, so that it can be adapted by any heterogeneous network systems. Though the performance measurement technique used in [96] is claimed to be adaptive with composite network services, actually it is difficult enough for the conflicting set of quality parameters. Again optimization of quality parameters that are inter-dependent is difficult with above said methodology.

SMI (Service Measurement Index) based cloud performance measurement of business services are introduced in [97]. This is claimed to be the first initiative for measuring business performances of service providers and their services based on some key performance indicators. There are a few QoS parameters listed that help to describe the QoS and the providers performance. A new modified view of the architecture is proposed here. On the top of the service catalogue (i.e. registries) a layer of monitoring is there. The responsibility of monitoring is dependent on the quality service and quantity measures of services. Another component in the layer is filter that actually works for threshold i.e. to how much extent services are compromised with their qualities. On the top of this layer there is the SMI broker layer. Broker basically takes the query from consumers and the required level of QoS. This layer has three duties. It negotiates Service Level Agreement (SLA) and actually decides the QoS deliverables of providers. Then it calculates the SMI score for services. Based on the scores the services are ranked. Among the top ranked service user choose the most suitable one. This layer addition in the architecture facilitates the framework for ready made quality calculations and efficient service selection. The representative quality metrics may vary from domain to domain. Thus, the list may be longer compared to the QoS metrics considered so far.

In [15], a QoE (Quality of Experience) dependent methodology is introduced that practically works via an interface named GRIA. The QoE takes the decision making part in selecting the services with the help of a recommender unit within the service broker. The recommender works in three phases. First it takes request from users and discovers services, then these services ranking are found (that are already done based on QoE). After that a it runs through a recommenders algorithm that decides the final score of the services and ultimately the broker (working as a recommender) recommends the services to users. The whole system works as a decision support system for service selection. The additional layer that implements the idea of recommendation is a key factor included within the architecture. Secondly the metrics of QoE can be far improved by talking more QoS related QoE parameters and using more efficient covariance calculations.

The work in [91] has incorporated federated broker that help in intermediating among several internal external services in cloud. It helps in cross service access for collaborative services. It is very important and one of the essential part of integrating services over cross domains. The vulnerability analysis done on the proposed architecture and the loopholes are identified with respect to its security issues. The QoS in case of cross domain service integration is a challenge yet to be solved.

Often there arises need of access pervasive service or integrate those services in future service clouds for higher functionalities. In [98] KASO middle ware subsystem works well for pervasive services in embedded network. There exists a KASO API through which KASO interacts with agents in environment. It also includes knowledge management services that include brokerage. This broker contains only agent manager and service request registry. This a completely agent based system and proposes the PRA (Perceptual Reasoning Agents) based model for managing pervasive services. The main lack here is the proper semantics in the knowledge manager with respect to service qualities. For resource provisioning services the broker is designed specially with the capabilities for handling various issues that may arise only in resource allocation and distribution services.

According to the work in [99] a resource broker is designed that works in the Collective layer of the designed architecture. Another similar kind of work is done in [100] that not only describe resource allocation scheme, it also discuss about the optimization done with respect to VM (Virtual Machine) placement over cross cloud environment. In this framework cloud broker has three units as cloud scheduler, deployment plan and a virtual infrastructure manager. The cloud scheduler has a optimized scheduling algorithm. It works upon several providers and takes their specifications as input and decides the optimal place. In deployment plan a set of VM templates are there depending on the scheduler output the plan works. The virtual infrastructure manager has interfaces adapted to different provider by with the migration is done. The key mathematics is under the integer programming paradigm. The broker in both the cases is highly platform dependent and sensitive to environment on which it has to be allocated. Specification of SLA will be different with respect to rest of the services. Thus it is a open area to contribute. Another open area is that the quality assessment and related billing and financial calculation and indicative parameter identification in case of immigration is very important.

Reference [101] works in the multi party collaborative environment specifically Grid environment for aggregating resource information. There is a newer concept of meta-broker that actually formed by accumulating some grid brokers. Each meta-broker interacts with other meta-brokers for information. It is assumed that per resource domain there exists one meta-broker. At the time of allocating resources all relevant meta-brokers are communicated by a final broker and based on some quality factors the best ranked broker is selected. Two resource allocation and ranking methods are discussed in [101]. The overhead and cost calculation is important idea to be carried out. The hierarchical distribution of brokers indeed helps to distinguish the responsibilities and making the brokers loosely coupled.

## 5.3 Proposed Broker Architecture for Service Provisioning

From the above discussions the components identified in broker architecture for service provisioning are Design time components, Runtime components, and other components. Design time components are the main blocks that are responsible for developing the application specific functionalities i.e. services, the specification about the non functionalities of those services, the data storing facilities, different types of networking within services of a particular domain. Runtime components are mostly the message passing protocol and standards specified. These components are discussed in the Fig. 7.

Brief discussion of each of the components are done as follows:

1. Design time components are identified as service information, non functional specification networking, etc. Service information is available in UDDI [1]. But the broker needs the service information to be structured efficiently. Service discovery or selection mechanism adapted by the broker will be using that service information. Non functional Specification of services plays important role in service provisioning. While offering a service service providers specify their level of

QoS and other non functional specification in their SLAs. Broker must have the capability of optimizing those specification while choosing services according to user requirements. Often users feedback has strong impact on this specification. Broker based approach for inclusion of feedback based selection is an important area to contribute. Service providers are connected by compliance information based on previous collaboration or organizational flexibility. The broker has to deal with provider negotiation and configuration management for end to end service provisioning. Thus the networking information among the service providers must be available to brokers.

2. SOAP based message passing is traditionally used in the domain of web services. The REST-ful web services are now very popular. The broker architecture should be able to accommodate these two types of message passing for federation.

3. Other than the above two there exist some decision making components like ranking services, quality assessment, quality of service calculation etc. discussed before. By other components the specialty of service such as routing, security issues, proper interoperability are handled. broker handles customer and providers from wider ranges. Thus some security mechanism must be there to safeguard the customer's identity. Authentication and access management is a different modules in broker. It monitors the user roles and provide the required services to the consumers. It takes the decision about whether to allow a service or not. Configuration checking and maintenance among the service providers business policies are handled by another module within broker.

Identification of such components is important for seamless service provisioning with cloud service brokers. Each of the identified component requires attention of further researches to provide a holistic brokerage solution.

# 6 Conclusion

Now we would provide a summary of our overall study. Some research Questions were raised at the Sect. 2. Table 5 gives the references which encountered the questions so far.

Considering all approaches so far we can summarize that the research Questions encountered by different approaches. Table 5 summarizes the questions from Sect. 2 and relevant number of approaches along with the reference.

In this review we have tried to view the service provisioning from different possible aspect. We have reflected the outcome of each aspects in the respective subsection. The coverage of important research questions is moderately validated by Fig. 8 and we are concluding the survey by identifying the future questions that are consolidated as follows

1. The services must be available with their complete meta data. But often in the domain of service provisioning this is one of the missing part. Existing service

**Table 5** References of approaches encountering research questions

| Question | No of approaches | References |
|----------|------------------|-----------|
| Q1 | 17 | [8, 15, 17, 23, 27, 32, 33, 41, 43, 46–48, 50, 51, 99, 101, 102] |
| Q2 | 14 | [6, 8, 12, 13, 16, 20, 26, 31, 35, 39, 48, 50, 98, 102] |
| Q3 | 8 | [11, 13, 33, 38, 42, 44, 96, 100] |
| Q4 | 13 | [7, 8, 10, 14, 17, 18, 20, 35, 45, 51, 53, 97, 102] |
| Q5 | 20 | [9, 27–45] |
| Q6 | 16 | [55–59, 61–70, 94] |
| Q7 | 8 | [15, 91, 96–101] |

**Fig. 8** Proportion of chosen references answering raised questions

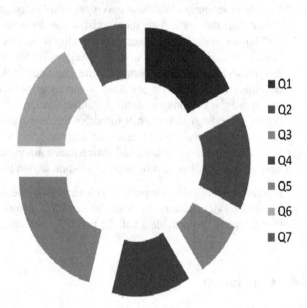

provisioning approaches can be modified to fulfill user requirements based on partial available information.

2. Cloud computing has a huge network dependent provisioning. There must be some standard mechanisms to measure network QoSs related to service performances. Some of these are identified in Sect. 3.2.

3. Dependency of services onto other services leads to the situations like Vendor Lock in. But to increase the customization efficiency and re usability the broker must be able to mitigate both the discovery and composition of services in a seamless manner avoiding such compliance and dependency issues.

4. Cost plays very important role in the service provisioning. Specially the service subscription based on Pay per use model in cloud demands a standardized fee structure and updated information of available resources. QoS and other

possible performance measure can be predicted and accordingly cost value may be attached to the services by broker.

5. Defining and measuring the QoS attributes and measure their aggregated effect on service provisioning is an important challenge of designing a broker.

6. Designing the security rules, Governing policies, and standardized instructions to vendors has prime importance. Moreover to safeguard users interest as well as provider's business interest; introduction of guaranteed service through the insuring [102] approach may be helpful in future.

**Acknowledgements** This publication is an outcome of the R&D work undertaken in the ITRA project of Media Lab Asia entitled "Remote Health: A Framework for Healthcare Services using Mobile and Sensor-Cloud Technologies".

# References

1. Liu, F., et al.: NIST cloud computing reference architecture. NIST Spec. Publ. **500**(2011), 292 (2011)
2. Cloud Service Broker Model-Sustainable Governance for Efficient Cloud Utilization: In: Lawler, C.M. (ed.) Green IT Cloud Summit 2012 Washington, D.C, April 18, Sheraton Premier, Tysons Corner
3. Samtani, G.: B2B Integration: A Practical Guide to Collaborative E-commerce. World Scientific (2002)
4. Yangui, S., et al.: CompatibleOne: the open source cloud broker. J. Grid Comput. **12**(1), 93–109 (2014)
5. Burt, J.: Gartner Predicts Rise of Cloud Service Broker-ages. http://www.eweek.com/c/a/Cloud-Computing/GartnerPredict-Rise-of-Cloud-Service-Brokerages-759833/
6. Garg, S.K., Versteeg, S., Buyya, R.: A framework for ranking of cloud computing services. Future Gener. Comput. Syst. **29**(4), 1012–1023 (2013)
7. Pawluk, P., et al.: Introducing STRATOS: a cloud broker service. In: 2012 IEEE Fifth International Conference on Cloud Computing. IEEE (2012)
8. Sundareswaran, S., Squicciarini A., Lin D.: A brokerage-based approach for cloud service selection. In: 2012 IEEE 5th International Conference on Cloud Computing (CLOUD). IEEE (2012)
9. Ferrer, A.J., et al.: OPTIMIS: a holistic approach to cloud service provisioning. Future Gener. Comput. Syst. **28**(1), 66–77 (2012)
10. World Wide Web consortium (W3C): Web Service Activity Statement. http://www.w3.org/2002/ws/Activity. Accessed 03 June 2007
11. Guillén, J., et al.: A service-oriented framework for developing cross cloud migratable software. J. Syst. Softw. **86**(9), 2294–2308 (2013)
12. Qu, L., Wang, Y., Orgun, M.A.: Cloud service selection based on the aggregation of user feedback and quantitative performance assessment. In: 2013 IEEE International Conference on Services Computing (SCC). IEEE (2013)
13. Villegas, D., et al.: Cloud federation in a layered service model. J. Comput. Syst. Sci. **78**(5), 1330–1344 (2012)
14. Cheng, D.-Y., et al.: A user centric service-oriented modeling approach. World Wide Web **14**(4), 431–459 (2011)
15. Tserpes, K., et al.: Service selection decision support in the Internet of services. In: Economics of Grids, Clouds, Systems, and Services, pp. 16–33. Springer, Berlin, Heidelberg (2010)
16. Wu, Q., et al.: A QoS-satisfied prediction model for cloud-service composition based on a hidden Markov model. Math. Probl. Eng. 2013 (2013)

17. Balan, R., Satyanarayanan, M., Park, S., Okoshi, T.: Tactics-based remote execution for mobile computing. In: Proceedings of the 1st International Conference on Mobile Systems, pp. 273–286. ACM, Applications and Services (2003)
18. Narayanan, D., Flinn, J., Satyanarayanan, M.: Using history to improve mobile application adaptation. In: Proceedings of Third IEEE Workshop on Mobile Computing Systems and Applications
19. Cuervo, E., Balasubramanian, A., Cho, D.-K., Wolman, A., Saroiu, S., Chandra, R., Bahl, P.: Maui: making smartphones last longer with code offload. In: Proceedings of the 8th International Conference on Mobile Systems, Applications, and Services, MobiSy10, pp. 49–62. ACM, New York, NY, USA (2010)
20. Flinn, J., Park, S., Satyanarayanan, M.: Balancing performance, energy, and quality in pervasive computing. In: Proceedings of the 22nd International Conference on Distributed Computing Systems, 2002, pp. 217–226. IEEE (2002)
21. Rochwerger, B., Breitgand, D., Levy, E., Galis, A., Nagin, K., Llorente, I.M., Galan, F.: The reservoir model and architecture for open federated cloud computing. IBM J. Res. Dev. 53(4), 4–1 (2009)
22. Zeng, C., Guo, X.A., Ou, W.J., Han, D.: Cloud computing service composition and search based on semantic. In: Jaatun, M.G., Zhao, G., Rong, C. (eds.) Cloud Computing, Proceedings, vol. 5931, pp. 290–300. Springer, Berlin (2009)
23. Liu, Y., Li, M., Wang, Q.: A novel user-preference-driven service selection strategy in cloud computing. Int. J. Adv. Comput. Technol. 4, 414–421 (2012)
24. Zhou, X., Mao, F.: A semantics web service composition approach based on cloud computing, pp. 807–810 (2012)
25. Zibin, Z., Xinmiao, W., Yilei, Z., Lyu, M.R., Jianmin, W.: QoS ranking prediction for cloud services. IEEE Trans. Parallel Distrib. Syst. 24, 1213–1222 (2013)
26. Paolucci, M., et al.: A broker for OWL-S web services. In: Extending Web Services Technologies, pp. 79–98. Springer, US (2004)
27. Usha, M., Akilandeswari, J., Syed Fiaz, A.S.: An efficient QoS framework for cloud brokerage services. In: 2012 International Symposium on Cloud and Services Computing (ISCOS). IEEE (2012)
28. Dastjerdi, A.V., Garg, S.K., Rana, O.F., Buyya, R.: CloudPick: a toolkit for QoS-aware service deployment across clouds. J. Autom. Softw. Eng. (2012)
29. Dutra, R.G., Martucci, M. Jr.: Dynamic adaptive middleware services for service selection in mobile ad-hoc networks. In: Mobile Wireless Middleware, Operating Systems, and Applications, pp. 189–202. Springer, Berlin, Heidelberg (2010)
30. Siebenhaar, M., et al.: Complex service provisioning in collaborative cloud markets. In: Towards a Service-Based Internet, pp. 88–99. Springer, Berlin, Heidelberg (2011)
31. Quarati, A., et al.: Hybrid clouds brokering: business opportunities, QoS and energy-saving issues. Simul. Model. Pract. Theory 39, 121–134 (2013)
32. Misra, S.C., Mondal, A.: Identification of a companys suitability for the adoption of cloud computing and modelling its corresponding Return on Investment. Math. Comput. Model. 53(3), 504–521 (2011)
33. Han, R., et al.: Enabling cost-aware and adaptive elasticity of multi-tier cloud applications. Future Gener. Comput. Syst. 32, 82–98 (2014)
34. Collazo-Mojica, X.J., Ejarque, J., Sadjadi, S.M., Badia, R.M.: Cloud application resource mapping and scaling based on monitoring of QoS constraints. In: Proceedings of the 2012 International Conference on Software Engineering and Knowledge Engineering, vol. 7, no. 4, pp. 88–93 (2012)
35. Li, W., et al.: Resource virtualization and service selection in cloud logistics. J. Netw. Comput. Appl. 36(6), 1696–1704 (2013)
36. Van den Bossche, R., Vanmechelen, K., Broeckhove, J.: Online cost-efficient scheduling of deadline-constrained workloads on hybrid clouds. Future Gener. Comput. Syst. 29(4), 973–985 (2013)

37. Devgan, M., Dhindsa, K.S.: QoS and Cost Aware Service Brokering Using Pattern Based Service Selection in Cloud Computing. Int. J. Soft Comput. Eng. **3** (2014)
38. Jula, A., Othman, Z., Sundararajan, E.: A hybrid imperialist competitive gravitational attraction search algorithm to optimize cloud service composition. In: 2013 IEEE Workshop on Memetic Computing (MC), pp. 37–43 (2013)
39. Zhao, X., Wen, Z., Li, X.: QoS-aware web service selection with negative selection algorithm. Knowl. Inf. Syst. **125** (2013)
40. Dou, W., Zhang, X., Liu, J., Chen, J.: HireSome-II: towards privacy-aware cross-cloud service composition for big data applications. IEEE Trans. Parallel Distrib. Syst. (2013)
41. Karim, R., Chen, D., Miri, A.: An end-to-end QoS mapping approach for cloud service selection. In: 2013 IEEE Ninth World Congress on Services (SERVICES), pp. 341–348 (2013)
42. Wang, S.G., Sun, Q.B., Zou, H., Yang, F.C.: Particle swarm optimization with skyline operator for fast cloud-based web service composition. Mob. Netw. Appl. **18**, 116121 (2013)
43. Ye, Z., Zhou, X., Bouguettaya, A.: Genetic algorithm based QoS-aware service compositions in cloud computing. In: Yu, J., Kim, M., Unland, R.: (eds.) Database Systems for Advanced Applications, vol. 6588, pp. 321–334. Springer, Berlin, Heidelberg (2011)
44. Zibin, Z., Xinmiao, W., Yilei, Z., Lyu, M.R., Jianmin, W.: QoS ranking prediction for cloud services. IEEE Trans. Parallel Distrib. Syst. **24**, 1213–1222 (2013)
45. Fei, T., Yuanjun, L., Lida, X., Lin, Z.: FC-PACO-RM: a parallel method for service composition optimal-selection in cloud manufacturing system. IEEE Trans. Ind. Inf. **9**, 2023–2033 (2013)
46. Li, Q., et al.: Model-based services convergence and multi-clouds integration. Comput. Ind. **64**(7), 813–832 (2013)
47. Van den Bossche, R., Vanmechelen, K., Broeckhove, J.: Cost-efficient scheduling heuristics for deadline constrained workloads on hybrid clouds. In: Proceedings of the 3rd IEEE International Conference on Cloud Computing Technology and Science, pp. 320–327. IEEE Computer Society (2011)
48. Kusic, D., Kandasamy, N.: Risk-aware limited lookahead control for dynamic resource provisioning in enterprise computing systems. In: Proceedings of the IEEE International Conference on Autonomic Computing, vol. 10, no. 3, p. 33750 (2010)
49. Anselmi, J., Ardagna, D., Cremonesi, P.: A QoS-based selection approach of autonomic grid services. In: Proceedings of the 2007 Workshop on Service-Oriented Computing Performance: Aspects, Issues, and Approaches, pp. 1–8. ACM, Monterey, California, USA (2007)
50. Kofler, K., Haq, I.U., Schikuta, E.: User-Centric, Heuristic Optimization of Service Composition in Clouds. LNCS, vol. 6271, pp. 405–417 (2010)
51. Kofler, K., ul Haq, I., Schikuta, E.: A parallel branch and bound algorithm for workflow QoS pptimization. In: ICPP '09. International Conference on Parallel Processing, 2009, pp. 478–485 (2009)
52. Moens, H., et al.: Cost-effective feature placement of customizable multi-tenant applications in the cloud. J. Netw. Syst. Manage. **22**(4), 517–558 (2014)
53. Hassan, M.M., Song, B., Huh, E.-N.: A market-oriented dynamic collaborative cloud services platform. Ann. Telecommun. (Annales des télécommunications) **65**(11–12), 669–688 (2010)
54. https://www.cloudfoundry.org/
55. https://cloudsleuth.net/
56. http://cloudharmony.com/
57. http://radlab.cs.berkeley.edu/wiki/Projects/Cloudstone
58. http://cloudcmp.net/
59. http://www.cloudclimate.com
60. http://www.cloudyn.com/
61. http://www.uptimesoftware.com/cloud-monitoring.php
62. http://cloudcruiser.com/
63. http://nagios.sourceforge.net/docs/nagioscore-3-en.pdf
64. http://opennebula.org/documentation:archives:rel2.0:img
65. https://github.com/zenoss/ZenPacks.zenoss.CloudStack

66. http://www.nimbusproject.org/
67. Chaves, S.A., Uriarte, R.B., Westphall, C.B.: Toward an architecture for monitoring private clouds. IEEE Commun. Mag. **49**, 130–137 (2011)
68. Corradi, A., Foschini, L., Povedano-Molina, J., Lopez-Soler, J.M.: DDS-enabled Cloud management support for fast task offloading. Comput. Commun
69. http://sourceforge.net/projects/hyperic-hq/
70. http://www.sonian.com/cloud-monitoring-sensu/
71. Abowd, G.D., Dey, A.K., Brown, P.J., Davies, N., Smith, M., Steggles, P.: Towards a better understanding of context and context-awareness. In: Proceedings of the 1st International Symposium on Handheld and Ubiquitous Computing, ser. HUC 99, pp. 304–307. Springer, London, UK. http://dl.acm.org/citation.cfm?id=647985.743843 (1999)
72. Abowd, G.D., Mynatt, E.D.: Charting past, present, and future research in ubiquitous computing. ACM Trans. Comput.-Hum. Interact. **7**, 29–58. http://doi.acm.org/10.1145/344949.344988 (2000)
73. Foldoc.org: Free on-line dictionary of computing. http://foldoc.org/context (2010). Accessed 21 May 2012
74. Verissimo, P., et al.: Cortex: Towards supporting autonomous and cooperating sentient entities, 595–601 (2002)
75. Hynes, G., Reynolds, V., Hauswirth, M.: A context lifecycle for web-based context management services. In: Barnaghi, P., Moessner, K., Presser, M., Meissner, S. (eds.) Smart Sensing and Context, ser. Lecture Notes in Computer Science, vol. 5741, pp. 51–65. Springer Berlin/Heidelberg. http://dx.doi.org/10.1007/978-3-642-04471-75 (2009)
76. Bellavista, P., Corradi, A., Fanelli, M., Foschini, L.: A survey of context data distribution for mobile ubiquitous systems, ACM Comput. Surv. **xx**(xx), 49 (2013). http://www-lia.deis.unibo.it/Staff/LucaFoschini/pdfDocs/contextsurveyCSUR.pdf
77. Strang, T., Linnhoff-Popien, C.: A context modeling survey. In: Workshop on Advanced Context Modelling, Reasoning and Management, UbiComp 2004—The Sixth International Conference on Ubiquitous Computing, Nottingham/England. http://elib.dlr.de/7444/1/Ubicomp2004ContextWSCameraReadyVersion.pdf (2004)
78. Casaleggio Associati: The evolution of internet of things, Casaleggio Associati, Technical Report, February 2011. http://www.casaleggio.it/pubblicazioni/Focusinternetofthingsv1.81. Accessed 08 June 2011
79. Peterson, M., Pierre, E.: Snias vision for information life cycle management (ilm), in Storage Networking World. Computer World (2004)
80. AIIM: What is enterprise content management (ecm)? February 2009. http://www.aiim.org/What-is-ECM-Enterprise-Content-Management.aspx. Accessed on 20 June 2012
81. Badidi, E., Esmahi, L.: A cloud-based approach for context information provisioning. arXiv preprint arXiv:1105.2213 (2011)
82. Falcarin, P., et al.: Context data management: an architectural framework for context-aware services. Serv. Oriented Comput. Appl. **7**(2), 151–168 (2013)
83. Gu, T., Pung, H.K., Zhang, D.Q.: A middleware for building context-aware mobile services. In: 2004 IEEE 59th Vehicular Technology Conference, 2004. VTC 2004-Spring, vol. 5. IEEE (2004)
84. Hofer, T., et al.: Context-awareness on mobile devices-the hydrogen approach. In: Proceedings of the 36th Annual Hawaii International Conference on System Sciences, 2003. IEEE (2003)
85. Object Management Group: The Common Object Request Broker (CORBA): Architecture and Specification. Object Management Group (1995)
86. Zhu, F., Mutka, M., Ni, L.: Splendor: a secure, private, and location-aware service discovery protocol supporting mobile services. In: Proceedings of the First IEEE International Conference on Pervasive Computing and Communications, 2003 (PerCom 2003). IEEE (2003)
87. Chen, T.: A fuzzy integer-nonlinear programming approach for creating a flexible just-in-time location-aware service in a mobile environment. Appl. Soft Comput. **38**, 805–816 (2016)

88. Xu, Y., et al.: Context-aware QoS prediction for web service recommendation and selection. Expert Syst. Appl. **53**, 75–86 (2016)
89. Wang, Y., et al.: CATrust: Context-Aware Trust Management for Service-Oriented Ad Hoc Networks (2016)
90. Anand, A., de Veciana, G.: Invited paper: context-aware schedulers: Realizing quality of service/experience trade-offs for heterogeneous traffic mixes. In: 2016 14th International Symposium on Modeling and Optimization in Mobile, Ad Hoc, and Wireless Networks (WiOpt). IEEE (2016)
91. Huang, H.Y., et al.: Identity federation broker for service cloud. In: 2010 International Conference on Service Sciences (ICSS). IEEE (2010)
92. Rasch, K., et al.: Context-driven personalized service discovery in pervasive environments. World Wide Web **14**(4), 295–319 (2011)
93. Jain, P., Rane, D., Patidar, S.: A novel cloud bursting brokerage and aggregation (CBBA) algorithm for multi cloud environment. In: 2012 Second International Conference on Advanced Computing & Communication Technologies (ACCT). IEEE (2012)
94. Lindner, M., et al.: The cloud supply chain: a framework for information, monitoring, accounting and billing. In: 2nd International ICST Conference on Cloud Computing (CloudComp 2010) (2010)
95. Simons, A.J.H., et al.: Advanced service brokerage capabilities as the catalyst for future cloud service ecosystems. In: Proceedings of the 2nd International Workshop on CrossCloud Systems. ACM (2014)
96. Duan, Q., Lu, E.: Network service description and discovery for the next generation internet. Int. J. Comput. Netw. (IJCN) **1**(1) (2009)
97. Garg, S.K., Versteeg, S., Buyya, R.: SMICloud: a framework for comparing and ranking cloud services. In: 2011 Fourth IEEE International Conference on Utility and Cloud Computing (UCC). IEEE (2011)
98. Corredor, I., Martínez, J.F., Familiar, M.S.: Bringing pervasive embedded networks to the service cloud: a lightweight middleware approach. J. Syst. Archit. **57**(10), 916–933 (2011)
99. Somasundaram, T.S., et al.: CARE Resource Broker: a framework for scheduling and supporting virtual resource management. Future Gener. Comput. Syst. **26**(3), 337–347 (2010)
100. Tordsson, J., et al.: Cloud brokering mechanisms for optimized placement of virtual machines across multiple providers. Future Gener. Comput. Syst. **28**(2), 358–367 (2012)
101. Rodero, I., et al.: Grid broker selection strategies using aggregated resource information. Future Gener. Comput. Syst. **26**(1), 72–86 (2010)
102. Bhattacharya, A., Choudhury, S.: Service insurance: a new approach in cloud brokerage. In: Applied Computation and Security Systems, pp. 39–52. Springer, India (2015)
103. Mokhtar, S.B., Preuveneers, D., Georgantas, N., Issarny, V., Berbers, Y.: Easy: efficient semantic service discovery in pervasive computing environments with qos and context support. J. Syst. Softw. **81**(5), 785808 (2008)
104. Hesselman, C., Tokmakoff, A., Pawar, P., Iacob, S., et al.: Discovery and composition of services for context-aware systems. Lect. Notes Comput. Sci. **4272**, 67 (2006)
105. Bellavista, P., Corradi, A., Montanari, R., Toninelli, A.: Context-aware semantic discovery for next generation mobile systems. IEEE Commun. Mag. **44**(9), 6271 (2006)

# Part II
# High Performance Computing

# Flexible Neural Trees—Parallel Learning on HPC

**Jiří Hanzelka and Jiří Dvorský**

**Abstract** The purpose of this research is to develop effective parallel Flexible Neural Tree learning algorithm based on Message Passing Interface at High Performance Computing environment. The implemented framework utilizes two bio-inspired evolutionary algorithms that were parallelized. Genetic algorithm is used to develop structure of FNT and differential evolution for fine tunning of the parameters. Framework was tested for its correctness and scalability on Anselm cluster. Scalability experiments prove good performance results.

**Keywords** Flexible neural tree · Artificial neural network · Parallel algorithm · MPI · HPC

## 1 Introduction

Artificial neural networks (ANN) have been successfully applied to a number of scientific and engineering areas in recent years. These areas include for example function approximation, time-series prediction, system control, image processing and so on. A performance of a neural network is highly dependent on its structure. A structure of ANN for particular problem is not usually unique, and there are a lot of ways how to define structure of ANN corresponding to the given problem. Depending on the problem there are many open questions how to construct the appropriate ANN—use only one or more hidden layers, use only feedforward or also feedback connections etc.

J. Hanzelka (✉) · J. Dvorský
Department of Computer Science, FEECS, VŠB – Technical University of Ostrava,
17. Listopadu 15, 708 33 Ostrava, Poruba, Czech Republic
e-mail: jiri.hanzelka@vsb.cz

J. Dvorský
e-mail: jiri.dvorsky@vsb.cz

© Springer Nature Singapore Pte Ltd. 2017                                                67
R. Chaki et al. (eds.), *Advanced Computing and Systems for Security*,
Advances in Intelligent Systems and Computing 568,
DOI 10.1007/978-981-10-3391-9_4

There were a number of attempts to design artificial neural network architecture automatically, such as constructive and pruning algorithms [1]. In 1997 Zhang et al. proposed concept of evolutionary induction of sparse neural tree [2]. Chen et al. in papers [3, 4] simplified concept of sparse neural tree to *Flexible Neural Tree* (FNT). In-depth description of the FNT can be found in book [5]. Peng et al. [6] proposed a parallel approach to evolving of FNT structure and its parameters using MPI. Probabilistic incremental program evolution (PIPE) method were used for structural optimizations and particle swarm optimization (PSO) was used for the parameters fine tuning.

The paper is organized as follows: Sect. 2 gives the brief description of the FNT. A parallel approach for evolving FNT is given in Sect. 3. Section 4 provides some experimental results for our implementation of FNT on HPC. Some concluding remarks are presented in Sect. 5.

## 2 Flexible Neural Tree

The Flexible Neural Tree is a special type of feed-forward artificial neural network. The neural network has an irregular tree structure with one output only. Formally speaking FNT is union of set of terminal symbols $T$ and set of functional symbols $F$ [5]:

$$Tree = F \bigcup T = \{+_1, +_2, \ldots, +_M\} \bigcup \{x_1, x_2, \ldots, x_K\} \tag{1}$$

The terminal symbols $x_1, \ldots, x_K$ are leaf nodes in the tree structure of the network and taking no arguments each. While functions symbols $+_1, \ldots, +_M$ are inner, non-leaf, nodes, taking $i$ input arguments. The functional symbol $+_i$ is also called a *flexible neuron operator* with $i$ inputs. Typical structure of the FNT is shown in Fig. 1.

During the FNT construction process, when a functional symbol $+_i$ is selected, $i$ real values $w_i$ are randomly generated and used as weights representing the connection strength between the symbol $+_i$ and its children. Moreover, two real numbers $a_i$ and $b_i$ are also randomly generated as adjustable parameters of activation function. The activation function is defined as:

$$f(a_i, b_i, z_i) = \exp - \left( \frac{z_i - a_i}{b_i} \right)^2, \tag{2}$$

where $z_i$ is the total excitation of $+_i$ given by:

$$z_i = \sum_{j=1}^{i} w_j x_j. \tag{3}$$

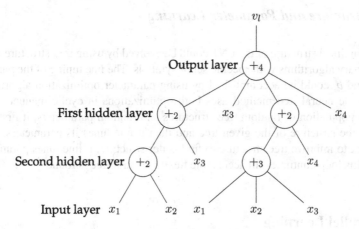

**Fig. 1** A typical representation of neural tree with functional symbols $F = \{+_2, +_3, +_4\}$, and terminal symbols $T = \{x_1, x_2, x_3, x_4\}$

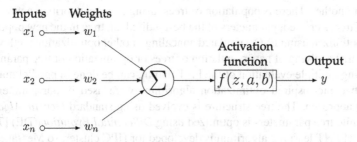

**Fig. 2** The total excitation and activation function of the flexible neuron $+_n$

The total excitation and activation function of the flexible neuron $+_n$ is illustrated in Fig. 2. The overall output of the network is computed recursively using depth-first traversal of the tree structure.

## 2.1 Fitness Function

A fitness function maps FNT to real number—fitness value. The fitness value represents the FNT performance on given task. The fitness function usually consists of two parts. The first part represents error measure i.e. MSE or RMSE between actual output data and FNT model output. The second part assesses structural properties of evaluated FNT, commonly number of nodes. If the two equal RMSEs are found, the FNT with smaller number of nodes is always preferred.

## 2.2  Structure and Parameter Learning

The hierarchical structure of the FNT could be evolved by using tree-structure based evolutionary algorithms with specific set of symbols. The fine tuning of the parameters $a_i$ and $b_i$ could be accomplished by using parameter optimization algorithms. The flexible neural tree method uses both optimizations in cyclic manner. Starting with population of random tree structures and random parameters, it first tries improve the structure of the given tree and then it fine tunes its parameters. Then goes back to improve tree structure to find better structure, it fine tunes parameters again. This loop continues till acceptable flexible neural tree is found.

# 3  Parallel Learning

The serial learning algorithm, see e.g. [5], consists of several steps that are executed one after another. There is population of trees, using genetic algorithm a new population of tree is created, parameters of the best individual, tree, in the new population is then optimized using PSO, simulated annealing or other optimization method. It is obvious that evolving of new population of trees or optimization of tree parameters, demanding multiple evaluation of the FNT output, can be done in parallel manner.

The two bio-inspired optimization algorithms were used in our parallel FNT learning approach. The tree structure is evolved using standard *Genetic Algorithm* (GA), while tree parameters is optimized using *Differential Evolution* (DE) [7].

Parallel FNT learning algorithm is developed for HPC cluster, so Message Passing Interface (MPI) [8–10] is straightforwardly used to implement GA and DE algorithms. After series of experiments, the GA was implemented using the asynchronous model where as the DE is implemented using the synchronous one. Asynchronous computation model for GA was selected due to very different sizes of FNT in the population and whence execution times for each tree in the population can very be different. Synchronous implementation of GA should wait for the longest execution over the largest FNT. While DE optimize only one given FNT, therefore execution time is the same for any set of FNT parameters.

General description of GA is given in Algorithm 1, master process in Algorithm 2 and slave processes in Algorithm 3. General description of the DE, master process and slave processes are given in Algorithms 4, 5 and 6 respectively.

---

**Algorithm 1** Parallel asynchronous genetic algorithm

---

Inputs: Population of FNTs, GA parameters
Outputs: New population of trees which is sorted, first FNT has best fitness

1. Parallel evaluate population of trees
2. Sort population by fitness, fitness is modified error where trees with same error but lower number of neurons have better fitness
3. Save best neural tree
4. Create new generation (by selection, crossover, reproduction, mutation)
5. Parallel evaluate new generation
6. Sort new generation by fitness
7. Repeat steps (3–5) until better neural tree than best tree is found or maximum generation step is not reached, otherwise go to step (8)
8. Return new population, first neural tree is best

---

---

**Algorithm 2** Parallel asynchronous genetic algorithm, *master process*

---

1. Send operation to evaluate individuals for async. GA to each Slave process by Broadcast MPI method
2. Sort population of flexible neural trees by number of neurons in descending order
3. Asynchronously receive message from any slave node
4. If received message is that Slave node wants work go to step (5) otherwise Slave node want to send result fitness so go to step (6)
5. If no more trees remain to evaluate then go to step (7) else send asynchronously to this Slave node neural tree to evaluate and go to step (3)
6. Asynchronously receive result fitness from this Slave node and go to step (3)
7. Send asynchronously to this Slave node end message, if this message was not send to all slave nodes go to step (3), else end evaluation

---

---

**Algorithm 3** Parallel asynchronous genetic algorithm, *slaver processes*

---

1. Wait for which operation to execute
2. If operation is evaluate individuals for async. GA

   (a) Send asynchronously message to Master process to get neural tree for evaluation
   (b) Asynchronously receive message from Master
   (c) If received message is neural tree then go to step (d), otherwise if it is end message go to step (1)
   (d) Evaluate fitness of received tree by root mean square error (RMSE) or other methods
   (e) Send asynchronously message to Master process about finished evaluation
   (f) Send asynchronously message to Master process with fitness of neural tree and then go to step (a)

---

**Algorithm 4** Parallel synchronous differential evolution

Inputs: FNT parameters, DE parameters
Outputs: New FNT parameters

1. Create random population of tree parameters and set input tree parameters to first individual
2. Parallel evaluate population
3. Find the best individual in population
4. Repeat steps (5 – 7) for each individual in population
5. Select 3 different parents, first is the best individual
6. Create noisy vector from parents
7. Create trial vector and save to trial population
8. Parallel evaluate trial population
9. Create new population by selecting better individual on same index from old population and trial population and find the new best individual
10. If fitness of the best individual is better than required fitness go to step (12) else go to step (11)
11. If maximum generation is reached then go to step (12), otherwise go to step (3)
12. Return tree parameters of the best individual

---

**Algorithm 5** Parallel synchronous differential evolution, *master process*

1. Send operation to evaluate individuals for DE to each Slave process by Broadcast MPI method
2. Send population by Scatter MPI method to all processes
3. Evaluate fitness of returned part of population for this process by root mean square error (RMSE) or other methods
4. Gather evaluated errors from all processes

---

**Algorithm 6** Parallel synchronous differential evolution, *slave processes*

1. Wait for which operation to execute
2. If operation is evaluate individuals for DE

   (a) Use Scatter MPI method to receive part of population for this process
   (b) Evaluate fitness of returned part of population for this process by root mean square error (RMSE) or other methods
   (c) Send back to Master process evaluated fitness values by Gather MPI method
   (d) Go back to step (1).

---

# 4 Experiments

**Experimental Hardware** Experiments were performed on Anselm supercomputer where each node had two 8-core processors Intel Sandy Bridge E5-2665 clocked at 2.4 GHz and 64 GB DDR3 1600 MHz memory. These nodes did not have GPU or MIC accelerators.[1]

---

[1] Detailed specification for Anselm can be found on the web https://docs.it4i.cz/anselm-cluster-documentation/hardware-overview.

**Table 1** Settings of FNT algorithm for prediction of Box-Jenkins series

| Parameter | Value |
|---|---|
| Max epoch steps | 10 |
| Required RMSE | 0.025 |
| Number of processes | 4 |
| GA population size | 256 |
| GA max generations | 100 |
| GA mutation rate | 0.2 |
| GA crossover rate | 0.2 |
| DE population size | 64 |
| DE iterations | 900 |
| DE crossover rate | 0.99 |

## 4.1 Time-Series Forecasting

Time-series forecasting is an important research and application area. The implemented FNT is tested on well-known time-series prediction benchmark problem—Box-Jenkins time series [11]. The gas furnace data (series J) was recorded from combustion process of a methane-air mixture. The data set consists of 296 pair of input-output measurement. The input $u(t)$ is the gas flow into the furnace and $y(t)$ is the $CO_2$ concentration in outlet gas. The sampling interval is 9 s.

The input of evolved FNTs are pairs $u(t-4)$ and $y(t-1)$ and the output value is $y(t)$. Values are then normalized to interval $\langle 0, 1 \rangle$. The input and output data are divided in two parts. First part is used for training (200 patterns) and the rest of the data is used for testing.

Parameters used in the test are shown in Table 1. The test was repeated 20 times. The best FNT evolved in each run was stored. The average RMSE value for these trees was 0.021 for training set and 0.047 for test set. The best FNT evolved in the test reach RMSE 0.019 for the training set and 0.038 for test test. The best FNT is shown in Fig. 4, where $x0$ and $x1$ denotes input variables $u(t-4)$ and $y(t-1)$, respectively. The actual time-series, the FNT output and the prediction error is given in Fig. 3. The experimental results are comparable to results obtained by Chen and Abraham [5], see Table 2.

## 4.2 Scalability

This test was carried out to check the scalability of proposed FNT parallel learning algorithm. The most time consuming part were evolution of FNT structure together with optimization of FNT parameters. The first evolution was done using genetic algorithm (GA) and the second one is done using differential evolution (DE), see Sect. 3. There are three possible ways of GA execution:

**Fig. 3** Comparison of desired output with actual output of founded neural tree for Box-Jenkins series

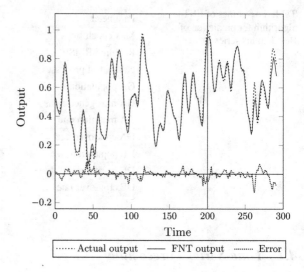

**Fig. 4** Flexible neural tree for prediction of Box-Jenkins time series

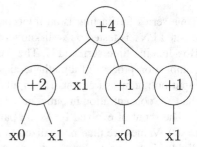

- serial,
- synchronous MPI and
- asynchronous MPI.

For each of above mentioned ways, execution times were measured and compared. The same time measurement was done for FNT parameters optimization done by DE.

Test was performed for 1000 and 5000 patterns in training set containing purpose-built artificial data. Both optimization algorithms used same population size with 128 individuals. Algorithm was tested for up to 64 cores (one process per core). FNT training algorithm was running for 10 epochs and speedup values obtained in each epoch are averaged. Speedup can be defined as

$$S = \frac{T_s}{T_p}, \tag{4}$$

where $T_s$ is execution time of serial algorithm and $T_p$ of parallel algorithm respectively. Speedup values is shown in Fig. 5.

**Table 2**  Comparative results of FNT approaches

| Model name | Number of inputs | Number of nodes | RMSE |
|---|---|---|---|
| FNT model (Case 1) [5] | 2 | 6 | 0.026 |
| Hanzelka & Dvorský | 2 | 4 | 0.038 |
| FNT model (Case 2) [5] | 7 | 3 | 0.017 |

**Fig. 5**  Speedup of optimization algorithms

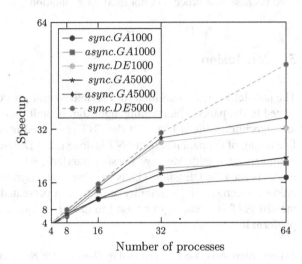

Per-process speedup can be defined as

$$S' = \frac{S}{n} = \frac{T_s}{T_p n},$$
(5)

**Table 3**  Average effectiveness of optimization algorithms

(a) 1000 patterns

| $S'$ | Number of processes | | | | |
|---|---|---|---|---|---|
| | 4 | 8 | 16 | 32 | 64 |
| Sync. GA | 0.86 | 0.75 | 0.69 | 0.48 | 0.27 |
| Async. GA | 0.71 | 0.79 | 0.85 | 0.63 | 0.34 |
| Sync. DE | 0.99 | 0.96 | 0.97 | 0.87 | 0.50 |

(b) 5000 patterns

| $S'$ | Number of processes | | | | |
|---|---|---|---|---|---|
| | 4 | 8 | 16 | 32 | 64 |
| Sync. GA | 0.88 | 0.83 | 0.70 | 0.59 | 0.37 |
| Async. GA | 0.74 | 0.86 | 0.91 | 0.92 | 0.55 |
| Sync. DE | 0.98 | 0.99 | 0.97 | 0.97 | 0.80 |

where $n$ is the number of computation processes. Per-process speedup represents contribution of each process participating in computation to whole speedup. In ideal case $S' = 1$ i.e. speedup is linear function of number of processes. In this case parallel algorithm scales linearly. Table 3a, b show that the averaged per-process speedup of parallel approaches increases with bigger training set. Per-process speedup of asynchronous GA approach increases over synchronous version when more processes are used because first process in not used for evaluation.

## 5 Conclusion

The parallel learning algorithm for Flexible Neural Tree on HPC platform was presented in the paper. The learning algorithm is built on two bio-inspired evolutionary algorithms. The structure of the FNT is evolved using Genetic Algorithm and fine tuning of the parameters of FNT is done using Differential Evolution. The FNT is implemented with Message Passing Interface, where synchronous computation model is used for DE whilst GA is running using asynchronous model. Implementation correctness and scalability are tested in presented experiments. The implemented FNT is intended to be used to model various scientific problems on HPC platform in general.

**Acknowledgements** Work is supported by Grant of SGS No. SP2016/68 and SP2016/97, VŠB - Technical University of Ostrava, Czech Republic.

## References

1. Nadal, J.P.: Study of a growth algorithm for a feedforward network. Int. J. Neural Syst. **1**, 55–59 (1989)
2. Zhang, B.T., Ohm, P., Mühlenbein, H.: Evolutionary Induction of Sparse Neural Trees. MIT Press (1997)
3. Chen, Y., Chen, F., Yang, J.Y.: Evolving MIMO Flexible Neural Trees for Nonlinear System Identification. CSREA Press (2007)
4. Chen, Y., Peng, L., Abraham, A.: Gene Expression Profiling Using Flexible Neural Trees. Springer, Berlin Heidelberg (2006)
5. Chen, Y., Abraham, A.: Tree-Structure based Hybrid Computational Intelligence: Theoretical Foundations and Applications. Springer (2010)
6. Peng, L., Yang, B., Zhang, L., Chen, Y.: A parallel evolving algorithm for flexible neural tree. Elsevier Science Publishers B. V. (2011)
7. Salam, A., Kader, A., Wahed, A.: Comparative study between differential evolution and particle swarm optimization algorithms in training of feed-forward neural network for stock price prediction. IEEE (2010)
8. Indiana University: MPI.NET: High-Performance C# Library for Message Passing (2015). http://www.osl.iu.edu/research/mpi.net/. Accessed 28 Feb 2015
9. Message Passing Interface Forum: MPI: a message-passing interface standard: Version 2.2 (2015). http://www.mpi-forum.org/docs/mpi-2.2/mpi22-report.pdf. Accessed 28 Feb 2015

10. Willcock, J., Lumsdaine, A., Robison, A.: Using MPI with C# and the common language infrastructure. ACM (2002)
11. Box, G., Jenkins, G.: Time Series Analysis: Forecasting and Control. Holden-Day, San Francisco (1970)

# Software Energy Estimation to Improve Power Efficiency of Embedded System

Nachiketa Chatterjee, Saakallya Biswas and Partha Pratim Das

**Abstract** Energy can be optimized for constrained-budget embedded system by using energy-aware processors and by techniques to minimize energy complexity in software coding. With sophisticated processor already in use, the latter is becoming the order of the day. For example, in Windows 8, a Battery Life Analyzer assists developers write energy-aware applications. In this paper, we focus on software energy optimization using simulation. We first develop a custom 8051 board to measure the energy consumed by a program (coded with a fixed set of instructions) excluding any additional overhead (of OS or monitor codes). We then estimate and trace the energy consumption of a software on this board and validate with an EFM32 Board. Based on these experimental data, we analyze different algorithms and data structures to identify factors to effectively improve energy consumption. Finally, we develop a simulator for energy estimation using PIN, a dynamic instrumentation framework by Intel. We validate the results of the simulator against those of the boards to suggest a simulation-based approach that can be developed into active assistance in a compiler for keeping software developers abreast of the energy needs.

**Keywords** Software energy optimization · Mobile · Embedded system

N. Chatterjee (✉)
A. K. Choudhury School of Information Technology,
University of Calcutta, Kolkata, India
e-mail: nachiketa.chatterjee@gmail.com

S. Biswas · P.P. Das
Department of Computer Science and Engineering,
Indian Institute Technology, Kharagpur, India
e-mail: riju91@gmail.com

P.P. Das
e-mail: ppd@cse.iitkgp.ernet.in

© Springer Nature Singapore Pte Ltd. 2017
R. Chaki et al. (eds.), *Advanced Computing and Systems for Security*,
Advances in Intelligent Systems and Computing 568,
DOI 10.1007/978-981-10-3391-9_5

# 1 Introduction

As mobile phones and tablets gain wide usage in every segment of daily life, and most stakeholders are now conscious and concerned about energy: end-users like Adams [1] realizes that certain applications can reduce battery life dramatically and considers energy consumption as an important quality attribute. Battery life is one of the important factors for mobile platforms [2]. In fact in a number of critical devices in Medical and Health monitoring, Civil Structure Health monitoring and Wireless Sensor Networks, we cannot really provide a continuous source of power and they need to be managed with a strict power budget. We can optimize at various levels and aim for a low energy consuming system. Following are the 3 main divisions where optimization research is being carried out:

- *Hardware Optimization* to develop energy efficient hardware for commonly used instructions to reduce average energy consumption, even optimize the idle time energy consumption.
- *Compiler Optimization* at the compiler level where instructions rescheduled and reshaped to reduce the transition density and enhance register usage, even generating platform dependent code to exploits their special hardware properties.
- *Software Optimization* to classify different algorithms based on their energy consumption and use the efficient one. We can isolate programming techniques which consume higher amount of energy and replace these with code snippets which use lesser amount of energy.

This paper mainly deals with Software energy optimization, where we have analyzed the correlation of time complexity and energy consumption with different other factors. Earlier a research paper [3] claimed few observation of energy consumption of different sorting algorithms for some of the embedded systems which is not aligned with our analysis. We prepared some experimental setup to identify the key factors of the energy consumption of different algorithms, which can help reduce the energy consumption of software.

# 2 Related Work

Some of the software energy optimization techniques was attempted earlier by use of following mechanism:

## 2.1 Resource Substitution

In case of bigger systems [4, 5] with various resources, time and energy consumption of a program depends on the weighted average of the time delay and energy

**Table 1** Resource substitution strategies

| Substitution | CPU | Communication | Memory |
|---|---|---|---|
| CPU | – | Migration of computation to server | Re-use of temporary memory |
| Communication | Calculation executed locally | – | Local storage of data |
| Memory | Data compression and compact data structure | Server carries out data management | – |

consumption of these components. If it is possible to accomplish same task using different resources, we can replace the resource. Three main aspects can characterize resources are utility, quantity, and use.

In resource substitution, usage of one of the resources is partially substituted by another resource, thus inhibiting over-usage and optimizing the usage of one resource. These different substitutions strategies are summarized in a compact form in Table 1.

## 2.2 Sources of Energy Optimization

In embedded systems, software optimization for energy is achieved by [6]: 1. Selection of the least expensive instruction or instruction sequences, 2. Minimizing the frequency of memory accesses, 3. Exploiting energy minimization features of hardware.

Usually the energy optimization objectives depend on the intended application. In battery powered systems, the total energy dissipation of a processing task determines how quickly the battery is spent. In systems where the power is constraint by heat dissipation and reliability considerations, instantaneous or average energy dissipation form important optimization objectives.

## 2.3 Instruction Selection and Ordering

Since there are many alternate code sequences that accomplish the same task, so it should be possible to select an optimized one in-terms of energy. Regarding performance and energy minimization, cache performance is a greater concern. Large code and small cache can lead to frequent cache misses with a high power penalty according to Roy et al. [6]. Also the concurrent operations of integer and floating point numbers can add to power conservation. Accumulator spilling and mode switching are sensitive to instruction ordering and are likely to have some energy impact. With a single accumulator, any time an operation writes to a new program variable, the previous accumulator value will need to be spilled to memory incurring the energy cost of a memory write.

## 2.4  Minimizing Memory Access Costs

*Since memory often represents a large fraction of system's energy budget, there is a clear motivation to minimize memory energy cost* [6]. Energy minimization techniques related to memory concentrate on the following objectives:

- Minimize the number of memory access required by an algorithm
- Minimize the total memory required by an algorithm
- Make memory access as close as possible to the processor: registers first, cache next and external RAM last.
- Make the most efficient use of the available memory bandwidth for example use multiple word parallel loads instead of single word loads.

## 2.5  Algorithm Selection

Bunse et al. published a paper on choosing the best sorting algorithm for optimal energy consumption [3] where they have built a system based on a micro-controller (Atmega 128, and external RAM running on a STK500/501 board) and found that the energy consumption and time complexity has very less correlation. So to optimize the energy consumption of an application we can chose the energy efficient algorithm. But in our observation energy consumption should go hand in hand with the time complexity as that depend on the number of instructions and the cost of those instructions along with a few other factors.

## 3  Objective

In this paper we intend to analyze the software energy consumption model and optimize it. Our specific objectives are as follows:

- To verify the observations of Bunse et al. [3] using different hardware setups and software simulation.
- To determine the power model with various instructions using:

  - 8051 based Board custom-designed for the purpose,
  - EFM32 Board as commercially available, and
  - Software simulation by dynamic binary instrumentation with PIN

- To validate the simulation results against the observations from the Boards.
- To estimate the energy consumption of different instruction with various data excluding the additional energy overhead due to different other factors like control of OS, IO, and caching etc.
- To identify factors controlling the power consumption of a program to provide assistance to the developer.

## 4    Experimental Setup

Our experimental setup will consist of 3 different parts:

## 4.1    Energy Calculation Using the 8051 Custom Board

This board as in Fig. 1 is purely fabricated from scratch to identify the energy consumption of a code fragment. We use AT89S52 [7], a low-power, high-performance CMOS 8-bit micro-controller, in our experimental setup. We know exactly what all components we are using without any abstraction. Even the operating system used is built from scratch and so are the application codes, so it is easier for us to analyze the results.

### 4.1.1    CPUPWR

The monitor code (Operating System) for the 8051 board was written in assembly which has the following features:

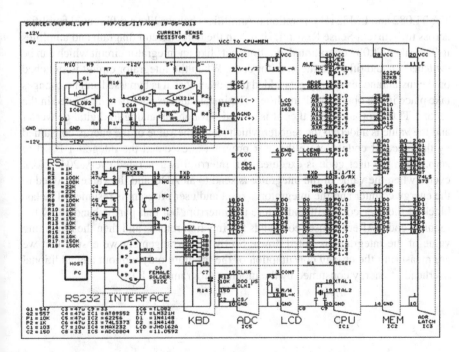

**Fig. 1**  Overview of custom board

- This code allows us to monitor a user program by showing us the energy consumed by a piece of user program and the number of clock cycles used by the program.
- We can download a set of data to the external RAM of the board.
- Similarly we can upload data from the external RAM to the computer.
- We have 6 slots for user programs where we can inject different codes and calculate the energy for each of them and analyze.
- The design is such that the code space is defined for each program for example 0x800 to 0xBFF for program 1 which has to be maintained by the user program by using ORG 0800H as the first line of the user program which sets the offset in the code memory.
- We can change the external RAM data from the board as well.
- The monitor code generates an interrupt every 2048 cycles when the user program is put on hold and the CPU is used by the monitor program to calculate the energy consumed during these 2048 cycles and that value is added to the global energy counter and the local energy counter is reset which calculates the energy consumed during these 20148 cycles. After that the user program is resumed. After execution of the user program we can view the energy log and the number of clock cycles taken by the user program.

### 4.1.2 Energy Calculation

The TL082 and LM321H Operational Amplifiers behave like a differential amplifier across the Current Sense Resistor, shown in the board circuit diagram and amplifies it. The other TL082 operational amplifier acts like an integrator circuit which can be switched on and off from the monitor code. When the integrator circuit is switched on the instantaneous energy gets added to it, otherwise it is bypassed and no changes take place in the operational amplifier. The instantaneous energy is integrated in the second TL082 op-amp and then it is passed through an analog to digital converter and the value is added to the register which stores the value of the energy consumed by the piece of code. Before the user program is executed the integrator is set to zero. Then the user code is executed. An interrupt is generated every 2048 clock cycles when the value of the integrator is send to an ADC and the value is added to the energy counter and the integrator is reset and user program is resumed. The last block of the user code will end before the interrupt but that part is handled by the monitor code. After the user code ends the function returns and then the remaining value of the integrator is noted down. So by this arrangement we ensure that we are measuring the energy consumption of the program only without any additional overhead of energy consumed by the monitor program and IO.

## 4.2 Energy Calculation Using EFM32 Board

To cross validate the experimental results we also taken a reference an industry made board where energy measurement is offered by the API of the board or there are ways to calculate the energy consumption which is tested beforehand.

We use Silicon Labs [8] EFM32G_STK3700 and Keil uVision (IDE) and Simplicity Studio (to interact with the board) [9], to verify the results observed in the custom board. Silicon Labs offer two important functions, the first one BSP_VoltageGet() returns the Voltage being supplied to the board and the second one BSP_CurrentGet() returns the instantaneous value of current flowing in the board. If we take the product of the two values returned by these function, we get the instantaneous energy consumption and summing over these values for all the instructions we get the total energy consumed by the program.

```
Function that calculates energy:
int energycalc() {
  float current, voltage;
  current = BSP_CurrentGet();
  voltage = BSP_VoltageGet();
  energycount = (int)(current*voltage)/1000;
  return energycount;
}
```

To calculate the energy consumption of a program, first we write the program to resemble the assembly code that will be generated by a program. As Keil IDE doesn't allow us to see or inject additional energy calculation code in the generated assembly code, so we write a program in such a manner that one line in the C program gets translated into one line of assembly. For example if we want to find the energy consumption of int c = A[i]; We write the code as:

```
int *d = A+i; Global_energy_counter +=energycalc();
int c = *d; Global_energy_counter+=energycalc();
```

## 4.3 Energy Estimation Using PIN Tool

We use dynamic binary instrumentation on different sorting programs to simulate the number of operations and sequence which depend on the program inputs. Pin [10] is a framework has a very rich API which has a lot of diverse functionality [11, 12].

We use the pintool to extract the number and type of instructions so that we can find the weighted sum of the energy consumption for each instruction and the cost of switching between instructions. The cost of each instruction and the instruction switching cost will vary from one setup to another so we have used a heuristic where we have taken cost of an ALU instruction as k and cost of a read instruction as 10 k and cost of a write instruction as 50 k. If these constants are changed we can get the energy estimation for other setups as well. We will take code fragments from the pintool source code to explain how the following objectives are realized:

- finding the number of instructions.
- finding the number of instructions in a particular function.
- finding the number of times a particular instruction type is executed for example how many times a mov or add instruction is executed.
- finding the above mentioned data for every function.
- getting a list of which instruction comes after which one and hence estimating the energy consumption of the instruction switches which basically happens because of the bit-flips in the registers. Linking the source code to the assembly and then pointing out which lines in the source code corresponds to which assembly lines and then finding the number of times different instructions are executed which will help us to find the energy hot-spots.

## 5 Observation and Results

We calculate the energy consumption of a sorting algorithm with different volume and ordering of data set. We also compare the results for two different boards.

### 5.1 Results for the Custom 8051 Board

Using bubble sort as example and we calculate the energy consumption with sorted and almost sorted data sets. Results for 8051 are tabulated in Table 2.

**Table 2** Energy consumption result of sorted almost-sorted data in 8051 custom board

| Data-size | Ordering | Cycles | Energy |
|-----------|----------|--------|--------|
| 32 | Almost sorted | 6793 | 635 |
| 32 | Sorted | 6531 | 634 |
| 64 | Almost sorted | 33710 | 3382 |
| 64 | Sorted | 25322 | 2533 |
| 128 | Almost sorted | 134161 | 13716 |
| 128 | Sorted | 99737 | 10132 |

**Table 3** Energy consumption result of sorted data in EFM32 board

| Data-size | Ordering | Energy |
|-----------|----------|--------|
| 32        | Sorted   | 4928   |
| 64        | Sorted   | 19548  |
| 128       | Sorted   | 81089  |

## 5.2 Results for the EFM32 Board

To verify the correctness of the result of 8051 board in Table 2 we compared the same algorithm and data set in EFM32 Board. We found the energy consumption for sorted data set as in Table 3. We find that the ratio of relative energy consumed for the EFM32 board to the custom 8051 board lies somewhere between 7.8 and 8 for sorted data, which suggests that the energy is being calculated properly as the 2 boards are in synch with each other. For the sorted data, we find that the ratio of energy consumed for $n = 32$ and $n = 64$ as in the ratio 1:4 and that is maintained for $n = 64$, $n = 128$ and the next one as well. So we can conclude that in the best case scenario of bubble sort the energy consumption is $O(n^2)$. Though this applies on the best case and in the best case swaps don't occur, only comparisons takes place, so we conclude that the energy calculation is done properly as in case of bubble if there is no memory write then time complexity and power complexity should go hand in hand. We also used the other benchmark algorithms with different data set.

## 5.3 Software Energy Estimation Using PIN

We find that the histogram obtained after comparing different sorting algorithms for input size $= 10$ shows that heap sort, merge sort consume more energy than bubble sort and insertion sort as in Fig. 2. At a first glance this might seem to contradict the fact that bubble sort and insertion sort algorithms take $O(n^2)$ time to compute the result while merge sort and heap sort take O(n log n) time to compute. This is not an anomaly because the overhead of these algorithms cost more than the computation cost like heap sort, that takes a lot of costly write operations to build the heap. One write operation in heap sort will nullify 50 CPU operations of bubble sort and insertion sort. For arrays of small size ($n = 10$) these writes cost more than the computation so energy consumption is more.

As the size of the array grows the computation cost increases as in Fig. 3 and the overhead is no longer the dominant factor and we find that bubble sort and insertion sort consume far more than merge sort, heap sort and quick sort when the size of the array is 1000 or more. Thus for sorting arrays of small size bubble sort and insertion sort cost less than the more algorithmically efficient sorting algorithms but as the size grows these algorithms tend to become highly inefficient.

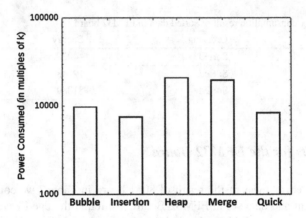

**Fig. 2** Energy consumption of different sorting with data size 10

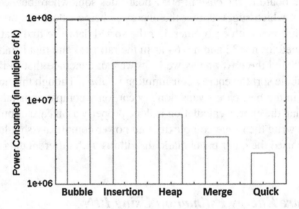

**Fig. 3** Energy consumption of different sorting with data size 1000

In Fig. 3 bubble sort and insertion sort consume energy of the order $(10^8 \, k)$ whereas the rest consume energy less than $(10^7 \, k)$. Since this is software energy estimation on a normal computer, we can accept that this might differ from the results obtained for embedded systems as there are a lot of differences between the two systems.

The presence of cache may contribute to the difference in observation and the boards which we have used later and those used by Bunse et al. [3] have 1 accumulator for general use so every computation that takes place is done using the following steps. First a value is loaded in the accumulator and then the computation is performed and it is saved in some register or memory location and then the next value is loaded in the accumulator so the presence of a single accumulator can act as a bottleneck. Hence we moved to the boards where we can calculate the energy consumption and which have similar setup. We have used a simple bubble sort code to check whether the two boards are giving similar results.

Isolate the parameters on which energy consumption depend and locate the code regions or programming techniques which are power hungry that we can avoid and use alternate methodologies.

Few experiments were conducted to check the parameters on which the energy consumptions of our setup depend. Firstly we took one instruction and looped over it numerous times with different sets of data to check whether we find any clue which may lead to isolation of various factors. The first set of data is taken for the program Power-set1:MOV A,R1; NOP; MOV A,R2; NOP;. The second set refers to the program where we do the MOVs and then do the NOPs keeping the number of cycles same, Power-set2:MOV A,R1; MOV A,R2; NOP; NOP;. The execution results plotted in Fig. 4. The graph clearly display the variation of energy consumption for different data set, i.e., the register values, where as in both the case the total number of cycle is 524814. We do the same thing for ALU operations like ADD

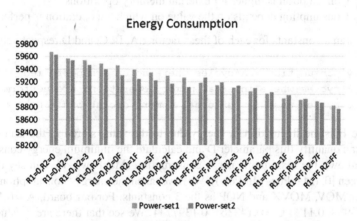

**Fig. 4** Energy consumption of 2 different instruction combination with different register value

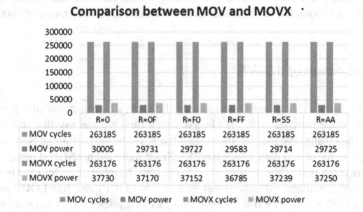

**Fig. 5** Energy consumption comparison between MOV and MOVX

and external memory operations like MOVX. Here we find that MOVX takes twice the number of cycles compared to internal memory operations or ALU operations and if we keep the number of cycles same, MOVX takes more energy, so basically 1 MOVX operation takes more than twice the amount of energy taken by MOV as in Fig. 5. We find that for almost the same amount of cycles MOVX takes almost 25% more energy and the value of the bits in the accumulator can result in a difference of 2.5% in the energy consumption. Similarly we took data for various instructions and we have found that the trend is somewhat as follows: Energy Consumption depends on the value stored in the accumulator for all the instructions. Energy Consumption is similar for internal memory operations and ALU operations. Now we have 4 factors which should be considered:

- Firstly if no operations are performed then some energy is consumed
- Energy consumption is lower for ALU or internal memory operations
- Energy Consumption is higher for external memory operations
- Energy Consumption depends on the value on which the operation is performed

We assign 4 constants for each of these factors: A, B, C, and D respectively.

```
Energy Consumption = A*cycles where no op are performed
    + B*cycles where internal memory operations are performed
    + C*cycles used for external memory operations + D
```

Where D is the fluctuation in energy due to the value stored in the accumulator. We can quantify this parameter D and calculate the minimum energy consumption (MinPow), the maximum energy consumption (MaxPow). We can say that D lies between [0, (MaxPow-MinPow)]. We have taken the energy "consumption/unit-cycle" for MOV, MOVX and NOP as the 3 constants. For our board, A = 0.11338 B = 0.113988, 0.1124 C = 0.143364, 0.1397734. We see that there are 2 values for B and C. That is because of the bit flips. If all the values are FF then we use the second constant. If all the values are 00 then we use the first constant. So the energy consumption is bounded by these values based on two different groups of data set.

## 5.4   Validation of Models

We need to verify this model by estimating the energy consumption of a program and if the value matches with the observed data then we can say that this model explains the results. We use the above equation to estimate the Bubble Sort Energy consumption for sorted data where n = 255. Since there are no NOP statements we can ignore the A part as the number of cycles in zero. Number of cycles used in internal memory operations = 4 outside any loop + 4 in LOOP1 and 6 in LOOP2 = 4 + 255 * 4 + (255 * 254/2) * 6 = 195334. Number of cycles used in external memory operations = 4 cycles outside loop + 0 cycles in loop1 + 4 cycles in loop 2 = 4 + (255 * 254/2) * 4 = 129544. So lower Bound on Total Energy = B * 195334 + C *

$129544 = 21955 + 18106 = 40061$ and upper bound on Total Energy $= B2 *$ $195334 + C2 *129544 = 22265 + 18571 = 40836$. Therefore the Energy Consumption should lie between (40061, 40836) and the observed data show that the energy consumption is 40286 which shows that the estimated energy bound is satisfied.

## 6 Conclusions

In this paper we replicate the experiment conducted by the Bunse et al. [3] by first trying with the 8051 custom board and measured the energy consumption of the program. Along with this board we took another board (EFM32) to verify the same. We found that the two boards were showing similar energy consumption differing by a constant factor.

We found that the energy consumption in an embedded system like a 8051 board depend mainly on the number of cycles used in the execution of the program. After inconclusive results we moved on to software energy estimation and developed a system to measure the energy consumption factors of the program.

We have also identified a few factors on which the power consumption of system like 8051 depends and we formed a model which can put a bound on the energy consumption from both sides. These bounds were tested and validated. We concluded that the power consumption of these systems depend on the value stored inside the register or accumulator and hence we can suggest changes in designs which will lead to a lower energy consumption. Op-codes can be designed in such a way that frequent operations have more number of 1's in their op-codes so that less amount of energy is wasted while loading them and with branch prediction mechanism we can put the more common branch address to be the one where more number of 1's are there. These techniques would lead to lower energy consumption and can be reduced by 2.5% without compromising on the time complexity. We have also observed that the external memory operations are very costly compared to the internal memory operations. It is advisable that while working with large data sets it is better to cache these data in the internal memory as external memory operations take much more than twice the energy. Dealing with bigger amount of data, caching in internal memory will create significant change in the energy consumption values.

The concepts developed may be further analyzed to build the entire power model and correlation between software energy consumption and the underlying hardware to improve the compiler optimization techniques to generate more energy efficient machine code in future.

# References

1. Adams, S.: Uncommunication Devices. http://dilbert.com/blog/entry/?EntryID=656. July 2011
2. Sabharwal, M.: Intel: Windows 8* Software Power Optimization, http://software.intel.com/en-us/articles/windows-8-software-power-optimization. July 16
3. Bunse, C., Hopfner, H., Roychoudhury, S., Mansour, E.: Choosing the "best" sorting algorithm for optimal energy consumption. In: Shishkov, B., Cordeiro, J., Ranchordas, A.K. (eds.) Proceedings of the 4th International Conference on Software and Data Technologie (ICSOFT 2009), 26–29 July 2009, Sofia, Bulgaria, vol. 2, pp. 199–206. INSTICC press, Setubal, Portugal (2009). ISBN: 978-989-674-010-8
4. Hopfner, H., Bunse, C.: Resource substitution for the realisation of mobile information systems. In: Proceedings of the 2nd International Conference on Software and Data Technologie. ICSOFT, pp. 283–289, (2007)
5. Hopfner, H., Bunse, C.: Towards an energy-consumption based complexity classification for resource substitution strategies. In: Proceedings of the 22nd Workshop on Foundations of Database(GVD), 25–28 May 2010, CEUR Workshop Proceeding, vol. 581 (2010)
6. Roy, K., Johnson, M.C.: Software Design for low power. http://www.cs.ucsb.edu/ckrintz/racelab/rre/papers/roy96software.pdf. Jul-16
7. AT89S52 Description: http://www.atmel.com/images/doc1919.pdf. July 16
8. EFM32 Giant Gecko Starter Kit: https://www.silabs.com/SupportDocuments/TechnicalDocs/efm32gg-stk3700-ug.pdf. July 16
9. Embedded Development Tools by ARM: http://www.keil.com/. July 16
10. PIN tool user guide: https://software.intel.com/sites/landingpage/pintool/docs/49306/Pin/html/. July 16
11. Luk, C-K., Cohn, R., Muth, R., Patil, H., Klauser, A., Lowney, G., Wallace, S., Reddi, V., Hazelwood, K.: Pin : Building Customized Program Analysis Tools with Dynamic Instrumentation
12. Hazelwood, K., Reddi, V.J.: Hands-On Pin for Architecture, Operating Systems, and Program Analysis Research

# Part III
# Image Processing

# Ternary Quantum Circuit for Color Image Representation

Sanjay Chakraborty, Sudhindu Bikash Mandal,
Soharab Hossain Shaikh and Lopamudra Dey

**Abstract** Image representation in a multilevel quantum system is always an important issue now a day. This paper initially proposes two approaches which help to represent color images in a ternary quantum system based on the modified concept of famous FRQI model and normalized amplitude based quantum representation model. But these approaches are complicated and have several drawbacks. Finally, a simple and a new model of color image representation and storage in a ternary (3-levels) quantum system is presented in this paper. This model deals with a set of quantum states for M different color levels and another set of quantum states for P different position coordinates. In this paper, various gray levels of a color image and their corresponding positions are stored in a $3^n$ color quantum register. For sake of simplicity this proposed method is carried out on $3 \times 3$ pixels of color image example and the model is built by using basic ternary gates. A basic measurement of a pixel in a quantum image is also presented in this paper. Comparisons among these three quantum image representation approaches are also discussed at the last section of this paper.

**Keywords** Multilevel quantum computing · Quantum color image processing · Ternary quantum image processing · Quantum information · Qubit · Qutrit

S. Chakraborty (✉) · S.B. Mandal
A.K. Choudhury School of IT, University of Calcutta, Kolkata, India
e-mail: schakraborty770@gmail.com

S.H. Shaikh
Computer Science and Engineering, BML Munjal University, Gurgaon, India

L. Dey
Computer Science and Engineering, Heritage Institute of Technology, Kolkata, India

© Springer Nature Singapore Pte Ltd. 2017
R. Chaki et al. (eds.), *Advanced Computing and Systems for Security*,
Advances in Intelligent Systems and Computing 568,
DOI 10.1007/978-981-10-3391-9_6

# 1 Introduction

To build futuristic computing systems, quantum computing plays a very important role in near future. It is a digital counterpart of quantum mechanics and quantum physics. After several developments like the integer factoring algorithm in polynomial time and quantum database search algorithm, quantum computing merge with the quantum image processing field to represent an emerging image processing technology by taking advantages of quantum computation. As for Quantum Image Processing, the research in the field has encountered fundamental difficulties. Several concepts of quantum image processing on a binary quantum system have been already proposed. Like famous polynomial based FRQI model for color image representation [1, 2] was first proposed. Inspired by this pioneering representation, numerous other quantum image representations have been suggested. They include a quantum mechanical approach for image retrieval and storage [3], a multi-channel quantum image representation based on phase transform (MCQI) [4], Caraiman's QIR approach [5], novel enhanced quantum representation of digital image model (NEQR), quantum image representation for log-polar images (QUALPI), simple quantum representation of infrared images (SQR) [6] and a normalized amplitude based quantum representation model [7]. Besides these, one quantum image processing technique has been introduced based on quantum entanglement [8]. However, all these ideas are formulated using the binary (2-level) quantum logic which is consistent with most approaches to quantum computing. The proposed approach refers to the mechanisms of representing and storing an RGB color image on a ternary (3-levels) quantum system. The advantage of using multilevel quantum system tells that the use of higher-dimensional quantum states increase the available Hilbert space exponentially with the same amount of physical resources. Besides that, an n-qutrit quantum system can be represented by a superposition of $3^n$ basis states, thus a quantum register of size n can hold $3^n$ values simultaneously (capable to store 24 bits RGB color image more effectively than binary system), whereas an n-qubits register can only hold $2^n$ values and also it uses more efficient ternary logical gates implementation. In particular, ternary quantum systems offer several benefits over binary quantum systems on representing, storing and processing of color images [9].

In this paper, a modified version of the famous FRQI model [1, 2] and a normalized amplitude based quantum representation model [7] are presented. These are suitable for color image representation in ternary quantum system. But these modified models are bounded by some typical drawbacks too. Like the complex modified FRQI model, opposing the postulates of quantum mechanics, the probability amplitudes of a quantum state cannot be accurately determined using a finite number of measurements. That is why the original classical image cannot be retrieved [10]. Besides that there are practical limitations on the number of colors/positions that can be physically represented using the angular values of the quantum phase of a qubit [10]. In particular, a basic ternary quantum logic circuitry is designed for representing color quantum images using some basic ternary quantum gates. This proposed approach uses the concept of M different color levels and P different position coordinates to represent color images by the help of $3^n$ color quantum register.

However, unlike the classical case, the same qubits are used to store the colors of all the pixels in the image. This is possible due to the principle of quantum superposition of states. This proposed approach help us to build the quantum circuit for color image representation in a simple way with less complexity, cost and effort. These approaches are discussed elaborately in later sections.

## 2 Proposed Models for Color Image Representation

### 2.1 First Approach Based on Modified FRQI Model

This approach is originated from the concept of the famous FRQI quantum image representation model [1–3]. The famous FRQI approach is only applicable for binary quantum systems. But this proposed approach is mainly used for representing color images on ternary quantum system. According to Flexible Representation of Quantum Images (FRQI) model, images on quantum computers can be represented in the form of a normalized state which captures information about colors ($|C>$) and their corresponding positions ($|P>$) in the images. It states that an image can be represented as,

$$|I(\theta)>= |C> \otimes |P> \tag{1}$$

$$|I(\theta)>= 1/2^n \sum_{i=0}^{2^{2n}-1} (\cos\theta_i|0> + \sin\theta_i|1>) \otimes |P> \tag{2}$$

where, $\theta_i \in [0, \pi]$, i= 0, 1, 2,..., $2^{2n} - 1$ and $\theta = \theta_0, \theta_1, \ldots, \theta_{2^{2n}-1}$ is the vector of angles encoding colors. There are two parts in the FRQI representation of an image; $\cos\theta_i|0> + \sin\theta_i|1>$ which encodes the information about colors and $|P>$ that about the corresponding positions in the image, respectively [2].

According to Kilmov qutrit phase model [11], the pure state of a qutrit is,

$$|\psi>= \sin(\frac{\xi}{2})\cos(\frac{\theta}{2})|0> + e^{i\phi_{01}}\sin(\frac{\xi}{2})\cos(\frac{\theta}{2})|1>$$
$$+ e^{i\phi_{02}}\cos(\frac{\xi}{2})|2> \tag{3}$$

where, $\theta$ and $\xi$ determine the magnitudes of the components of $|\phi>$, while we can interpret $\phi_{01}$ as the phase of $|0>$ relative to $|1>$ and analogously for $\phi_{02}$ and $\alpha = \sin(\frac{\xi}{2})\cos(\frac{\theta}{2})$, $\beta = e^{i\phi_{01}}\sin(\frac{\xi}{2})\cos(\frac{\theta}{2})$ and $\gamma = e^{i\phi_{02}}\cos(\frac{\xi}{2})$, where $|\alpha|^2 + |\beta|^2 + |\gamma|^2 = 1$. In case of qutrits based quantum system, it can be represented as,

$$|I(\theta)>= 1/3^n (\sum_{i=0}^{3^{2n}-1} \sin(\frac{\xi}{2})\cos(\frac{\theta}{2})|0> + e^{i\phi_{01}}\sin(\frac{\xi}{2})$$
$$\cos(\frac{\theta}{2})|1> + e^{i\phi_{02}}\cos(\frac{\xi}{2})|2>) \otimes |P> \tag{4}$$

Let take an example of a $3 \times 3$ color image (For sake of simplicity, we consider 9 different colors or different shades of RGB plane) and also there are total 9 positions, $|P_1> = |00>, |P_2> = |01>, |P_3> = |02>, |P_4> = |10>,$ $|P_5> = |11>, |P_6> = |12>, |P_7> = |20>, |P_8> = |21>, |P_9> = |22>$.
According to FRQI model, we can solve the angle value $\xi$ here by reading the color frequency of electromagnetic wave. And we can do that by the formula, Create a bijective function from a color to an angle $F1 : color \leftrightarrow \phi$, where $color = color_1, color_2, color_3, \ldots, color_M$, $color_i$ corresponds to the $i^{th}$ color in M colors, $\xi = \xi_1, \xi_2, \xi_3, \ldots, \xi_M$ [1],

$$\xi_i = \frac{\pi(i-1)}{2(M-1)} \tag{5}$$

where, $i \in (1, 2, 3, \ldots, M)$ If we put the value of $\xi$ within cos function or sin function, it will give a range of values between 0 and 1. This is the probability for which we can assign different colors to individual color pixels $|C>$. The state which occupies highest probability the pixel represents that particular color. (Let consider, $|0>$ represents Red, $|1>$ represents Green, $|2>$ represents Blue in RGB color plane). This is represented by Fig. 6.

RGB color image needs 24 bits i.e. $2^{24}$ color combinations. So in corresponding quantum model, it needs nearly $3^{16}$ basis states in an n-qutrit quantum register. So using this color quantum register, 9 different grey levels of the $3 \times 3$ color image can be stored with different probability of $\alpha$. Like, when $|P_1> = |00> = (00000000)^T$, then suppose $\alpha_1 = 0.7(MAXIMUM) = |C1> \leftarrow Red$ Color. Count all values in position bits are zero. $|P_2> = |01> = (01000000)^T$, then suppose $\alpha_2 = 0.2(MAXIMUM) = |C2> \leftarrow Green$ Color. Count the second left most bit position is 1 and others are 0. This same technique continues for other positions.

Now return to the first concept, suppose a particular pixel contains Red color, then we will have to represent that Red color in $|C1>$ by the help of some binary values like if $\alpha_1 = 0.7$ then the 7th bit value of the color pixel represents 1, others are $0.|C1> = 00000010$, now if we can apply Z-permutation gate operations 8 times on $|C1>$, like Z(+1), then that '1' color bit values shifted total 8 positions and each and every position holds or represents 'Various shades (intensities) of Red colors'. The similar logic is also applied for Green and Blue color variations as well. In case of Green value, if $\alpha_2 = 0.7$ then the 7th and successor 8th bits values of the color pixel represent 1, others are $0.|C2> = 00000011$, then again apply Z(+1) operations on them. So, when we retrieve the image pixel of that particular position, then we first count the location of 1 in the kronecker product result of $|02>$ and then we assign the value of its corresponding $|\alpha_i>$ and it fetches the corresponding stored color.

## 2.2 Second Approach Based on Modified Image Amplitude Normalization Technique

Srivastava et al. has introduced an idea to represent a color image using 2D quantum states and normalized amplitudes, which is essentially a binary quantum system

[7]. In this paper, this concept is used to represent color images for ternary quantum system. Unlike FRQI model, this approach is suitable for color images of any dimension. The storage space for the proposed method would only depend on the image dimensions [7]. The concept of dual representation of a 2-D image by row-location and column location vectors is used here. To represent a pixel location in its 2-D matrix, the tensor product of row-location and column-location vectors are used,

$$L_{p,q} = |I>_p \otimes <J|_q \qquad (6)$$

where,

$$|I>_p = |i> \otimes^m \qquad (7)$$

and

$$<J|_q = |j> \otimes^n \qquad (8)$$

where, $|I>_p$ is the row-location vector or the state of $m - qutrit$, $i \in 0, 1, 2$, $m = log_2 M$ and p is the row number of pixel and $<J|_q$ is the column location vector or the state of $n - qutrit$, $j \in 0, 1, 2$, $n = log_2 N$ and q is the row number of pixel. $L_{p,q}$ is the 2-D quantum state of a pixel at pth row and qth column using $m - qubits$ and $n - qubits$, respectively (Suppose we have $M - length$ row location vector with m qutrits and $N - length$ column location vector with n qutrits).

To illustrate with an example, a pixel location in terms of 2-D quantum states using row-location and column-location vector for any $3 \times 3$ image matrix is given by,

$$L_{p,q}^{3\times3} = \begin{bmatrix} |00> \otimes <00| & |00> \otimes <01| & |00> \otimes <02| \\ |01> \otimes <00| & |01> \otimes <01| & |01> \otimes <02| \\ |02> \otimes <00| & |02> \otimes <01| & |02> \otimes <02| \end{bmatrix}$$

where, 2-qutrit vector is obtained from the single qutrit constituent vectors as follows,

$$|01> = |0> \otimes |1> = \begin{bmatrix} 1 \\ 0 \\ 0 \end{bmatrix} \otimes \begin{bmatrix} 0 \\ 1 \\ 0 \end{bmatrix} = (010000000)^T$$

$$|12> = |1> \otimes |2> = \begin{bmatrix} 0 \\ 1 \\ 0 \end{bmatrix} \otimes \begin{bmatrix} 0 \\ 0 \\ 1 \end{bmatrix} = (000001000)^T$$

Similarly, we can follow the same representation for other row-location vectors. And for the column location vector it can be represented as, $< 01| = (010000000)$, $< 12| = (000001000)$ etc. Let $A_{p,q}$ be the amplitude/intensity of the pixel at pth row and qth column and $\alpha_{p,q}$ is the scalar amplitude of the pixel quantum state at pth row and qth column. $\alpha_{p,q}$ can be written as [7],

**Table 1**  A $3 \times 3$ color (RGB) image with its intensity values

| $A_{1,1}$ | $A_{1,2}$ | $A_{1,3}$ |
|---|---|---|
| (124, 117, 205) | (224, 115, 217) | (154, 187, 210) |
| $A_{2,1}$ | $A_{2,2}$ | $A_{2,3}$ |
| (224, 217, 145) | (96, 120, 220) | (175, 87, 100) |
| $A_{3,1}$ | $A_{3,2}$ | $A_{3,3}$ |
| (230, 130, 75) | (250, 125, 215) | (128, 99, 199) |

$$\alpha_{p,q} = \frac{\sqrt{A_{p,q}}}{\sqrt{A_T^{MXN}}} \tag{9}$$

where, $A_T^{MXN} = \sum_{p=1}^{M} \sum_{q=1}^{N} A_{p,q}$ As for example, the scalar amplitudes for 2-D quantum state for each pixel in images with $3 \times 3$ dimension is,

$$\alpha_{p,q}^{3 \times 3} = \frac{1}{\sqrt{A_T^{3 \times 3}}} \begin{bmatrix} \sqrt{A_{1,1}} & \sqrt{A_{1,2}} & \sqrt{A_{1,3}} \\ \sqrt{A_{2,1}} & \sqrt{A_{2,2}} & \sqrt{A_{2,3}} \\ \sqrt{A_{3,1}} & \sqrt{A_{3,2}} & \sqrt{A_{3,3}} \end{bmatrix}$$

Now we incorporate these structures for complete representation as an image. Making the use the 2-D quantum state representation of each pixel and its scalar amplitude, the quantum images can be represented as the superposition of all pixel's quantum states along with their scalar amplitudes. The proposed quantum image($Y_{p,q}$) representation is defined as (Table 1),

$$Y_{p,q} = \sum_{p=1}^{M} \sum_{q=1}^{N} \alpha_{p,q}^{3 \times 3} L_{p,q} \tag{10}$$

## 2.3  Proposed Approach Based on Basic Ternary Logic Circuitry

Ternary quantum logic is a qutrit based logic where more than two quantum basis states are introduced, for instance ($|0>, |1>, |2>$). This approach is simply based on reversible ternary circuitry logic. A ternary logic circuit can be made by the help of several ternary (qutrit) logic gates, like basic permutative gates, ternary Feynman gate, Muthukrishnan-Stroud (M-S) gate, ternary Toffoli gate, generalized ternary gate (GTG), Chrestenson (CH) gates, S-gate etc. The detailed discussion of all these

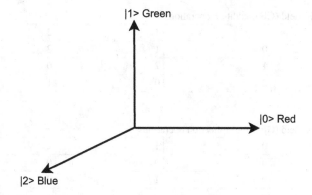

**Fig. 1** Three dimensional representation of RGB color range

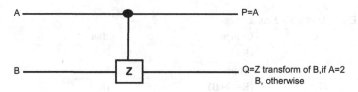

**Fig. 2** Graphical representation of Muthukrishnan-Stroud (M-S) Gate

ternary logic gates are described in several works [12–16]. To build the logic circuit for representing color images in the proposed approach, we mainly need the help of ternary toffoli gate, permutative gates and M-S gate with Galois field operations. Ternary Quantum Logic is the simplest introduction of multi-valued logic which is also referred to as 3VL. To define ternary logic, let T = (0, 1, 2). A ternary reversible logic circuit with n inputs and n outputs is also called an n-qudit ternary reversible gate can operate on ternary values [17]. In ternary quantum logic, the Galois-field algebraic structure will be the fundamental for constructing a unified approach to multiple-valued quantum logic [13]. The necessary ternary gates which are used to build the proposed circuit are described in Figs. 1 and 2.

### 2.3.1 Ternary Galois Field (GF3) Operations and M-S GATE

It consists of the set of elements T = 0, 1, 2 and two basic binary operations, addition (denoted by +) and multiplication (denoted by dot or absence of any operator) are shown in Tables 2 and 3. From this concept of GF3 logic, one famous gate is introduced (Muthukrishnan-Stroud or M-S gate) as follows,

**Table 2**  Galois Field (GF3) addition operations

| + | 0 | 1 | 2 |
|---|---|---|---|
| 0 | 0 | 1 | 2 |
| 1 | 1 | 2 | 0 |
| 2 | 2 | 0 | 1 |

**Table 3**  Galois Field (GF3) multiplication operations

| . | 0 | 1 | 2 |
|---|---|---|---|
| 0 | 0 | 0 | 0 |
| 1 | 0 | 1 | 2 |
| 2 | 0 | 2 | 1 |

**Table 4**  Example of a $3 \times 3$ color image

| Red | Orange | Blue |
|---|---|---|
| (R) | (R+G) | (B) |
| Cyan | White | Pink |
| (B+G) | (R+G+B) | (R") |
| Green | Light green | Black |
| (G) | (G") | |

### 2.3.2  Ternary Toffoli Gate

The basic operations of a ternary 3-qutrit Toffoli gate in the universal ternary quantum get set is shown in Fig. 3.

### 2.3.3  Proposed Methodology

In this approach, a RGB color image can be represented as 24 bit of color values and 2 bits of position values (in case of $3 \times 3$ color image). This allows storing an image with $N = 3^n \times 3^n$ pixels. As because a 3-level quantum register can hold maximum of $3^n$ basis states(by superposition), it needs total $3^3$ basis states to represent a RGB color image. The detailed representation of this approach on a $3 \times 3$ color image is described in Table 4.

Where, R" and G" refers to the variations of Red and Green colors. The equivalent ternary reversible circuit to represent the first co-ordinate of the $3 \times 3$ color image is shown in Fig. 4 with active-2 output [cost $= 14$], where the position bits are $|P_0 > |P_1 >= 00$, color bits $|CR_1 > \ldots |CR_8 >$ are 1 and all other color bits $|CB_1 > \ldots |CB_8 >$ and $|CG_1 > \ldots |CG_8 >$ are 0.

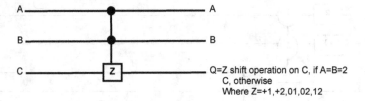

**Fig. 3**  Graphical representation of ternary Toffoli Gate

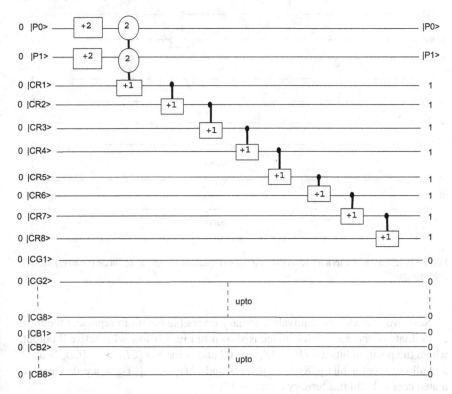

**Fig. 4**  Realization of a ternary reversible circuit to represent first co-ordinate (*Red*) of $3 \times 3$ color image

The equivalent ternary reversible circuit to represent the second co-ordinate of the $3 \times 3$ color image is shown in Fig. 5 with active-2 output [cost = 10], where the position bits are $|P_0 > |P_1 >= 01$ and color bits consist of a certain combination of $|CR_1 > \ldots |CR_8 >$.

Now, we consider the equivalent ternary reversible circuit to represent the seventh co-ordinate of the $3 \times 3$ color image is shown in Fig. 6 below with active-2 output, where the position bits are $|P_0 > |P_1 >= 20$ and color bits $|CG_1 > \ldots |CG_8 >$ are 1 and all other color bits $|CR_1 > \ldots |CR_8 >$ and $|CB_1 > \ldots |CB_8 >$ are 0. The estimated cost to build this ternary circuit is 14 (Figs. 5 and 6).

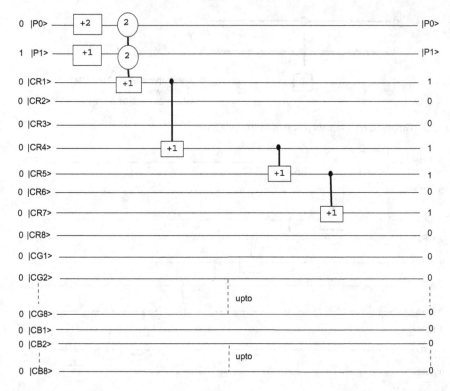

**Fig. 5** Realization of a ternary reversible circuit to represent second co-ordinate (*Orange*) of $3 \times 3$ color image

Next, we consider the equivalent ternary reversible circuit to represent the third co-ordinate of the $3 \times 3$ color image is shown in Fig. 7 below with active-2 output, where the position bits are $|P_0 > |P_1 >= 02$ and color bits $|CB_1 > \ldots |CB_8 >$ are 1 and all other color bits $|CR_1 > \ldots |CR_8 >$ and $|CG_1 > \ldots |CG_8 >$ are 0. The estimated cost to build this ternary circuit is 14.

## 3 Measurement of a Pixel's Quantum State

A modified version of quantum image amplitude normalization technique is mainly focused in this paper. In the above section, a proposed ternary logic circuitry is shown to effectively store the color and positions of pixels in an image. These quantum images can be retrieved using projective quantum measurements or POVM techniques without collapsing the quantum states [1, 5, 9]. In this approach a quantum state of a 2D quantum image can be represented as,

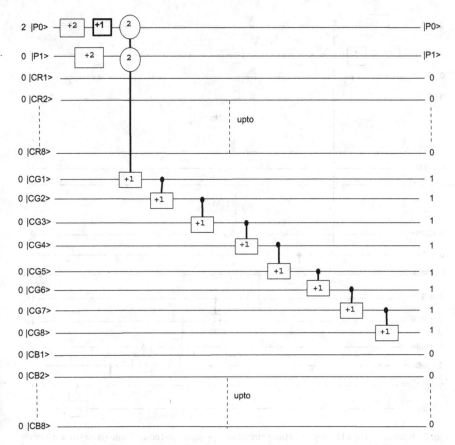

**Fig. 6** Realization of a ternary reversible circuit to represent seventh co-ordinate (*Green*) of 3×3 color image

$$Y_{p,q} = \frac{1}{\sqrt{A_T^{MXN}}} \sum_{p=1}^{M} \sum_{q=1}^{N} \sqrt{A_{p,q}}(|I>_p \otimes <J|_q) \qquad (11)$$

Here, two quantum registers are considered to store m ($log_2M$) and n ($log_2N$) qubits in the state,

$$|\psi> = \sum_{i=0}^{2^m-1} \sum_{j=0}^{2^n-1} \alpha_{i,j}|i,j> \text{ with } \sum_{i,j} \alpha_{i,j}^* \alpha_{i,j} = 1 \qquad (12)$$

The base-vectors $|i,j\rangle$ are interpreted as a pair of binary numbers with $i < 2^n$ and $j < 2^m$. The probability $p(I)$ to measure the number $I$ in the first register and the according post measurement state $|\psi_I'\rangle$ are given by,

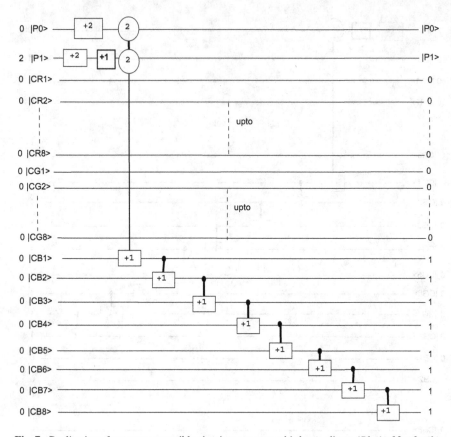

**Fig. 7** Realization of a ternary reversible circuit to represent third co-ordinate (*Blue*) of $3 \times 3$ color image

$$p(I) = \sum_{j=0}^{2^m-1} c_{I,j}^* c_{I,j} \tag{13}$$

and

$$|\psi_I'> = \frac{1}{\sqrt{p(I)}} \sum_{j=0}^{2^m-1} c_{I,j}|I,j> \tag{14}$$

If this modified ternary amplitude normalization technique is used to store and represent a color pixel of an image, then this method does not require any additional qubits to store it's 2D quantum state. It is totally independent of its bit depth. But FRQI approach used additional qubit to represent pixel amplitude. So, the first one is also less prone to decoherence as decoherence in quantum system is highly dependent upon the number of qubits [7].

In similar way, the different shades of primary red, green and blue color for different positions of the $3 \times 3$ color image can be represented by the help of ternary reversible circuit. A detailed comparisons among the above three approaches are given below,

## 4   Comparisons Among Above Three Quantum Image Representation Approaches

1. In FRQI model, to encode color information of a pixel the probability amplitudes of the corresponding one qubit state is required. According to the postulates of quantum mechanics, the probability amplitudes of a quantum state cannot be accurately defined using a finite number of measurements. Only simple color pixels operations are possible here due to the use of a single qubit state. In normalized amplitude based approach, the number of qubits increases with the image dimension while storing an image. So extra storage space is required to store information per pixel. One extra fractional bit is required to represent row-location vector or column-location vector. In proposed ternary based approach, it needs n qubits to represent $L = 2^n$ color values. So unlike FRQI model (need one qubit) it occupy more memory space to represent color values of an image. It uses n qubit states rather than one qubit state to represent an image.
2. Complex quantum circuit is required for implementation of this FRQI approach. Less complex circuit is required to implement our proposed ternary based approach.
3. The angular values of the quantum phase are not quantified in FRQI approach. So, it is difficult to identify which angle represents which color. Beside this, there is a practical limitation to physically represent angular values of the quantum phase of a qubit. No concept of these angular values of the quantum phase is required in our proposed ternary based approach.
4. This proposed approach requires (m+n) qubits to store $2^m \times 2^n$ dimension real type of image, where as FRQI approach requires (2m+1) qubits to store $2^m \times 2^m$ dimension real type of image. So, FRQI approach is suitable for square images but the proposed approach is suitable for all types of images. Also, Also, if the number of qubits are increased then the quantum states are more prone to decoherence.

## 5   Conclusion and Future Work

This paper is a combination of three different approaches for representing color quantum images in ternary quantum system. The main aim of this paper is to establish such a simple approach which provides a basis for the quantum color image representation operation using basic qutrit gates in a ternary quantum system. However, the proposed modified ternary approach will satisfy the famous Holevo's theorem which can give an upper bound to the amount of information which can be known about a quantum state. The future work will focus on developing various quantum image processing operations like, color image segmentation, image compression, image histogram technique etc. using row-location and column-location quantum states and operations for various image applications.

# References

1. Li, H.-S., Qingxin, Z., Lan, S., Shen, C.-Y., Zhou, R., Mo, J.: Image storage, retrieval, compression and segmentation in a quantum system. Quantum Inf. Process. **12**(6), 2269–2290 (2013)
2. Le, P.Q., Dong, F., Hirota, K.: A flexible representation of quantum images for polynomial preparation, image compression, and processing operations. Quantum Inf. Process. **10**(1), 63–84 (2011)
3. Verma, A.: Quantum image storage, retrieval and teleportation. Int. J. Adv. Res. Comput. Sci. Softw. Eng. **3**(10), 387–391 (2013)
4. Song, X.-H., Wang, S., Niu, X.-M.: Multi-channel quantum image representation based on phase transform and elementary transformations. J. Inf. Hiding Multimedia Signal Process., (**5**)4 (2014)
5. Caraiman, S., Manta, V.I.: Image segmentation on a quantum computer. Quantum Inf. Process. 1693–1715 (2015). Springer
6. Yan, F., Iliyasu, A.M., Venegas-Andraca, S.E.: A survey of quantum image representations. Quantum Inf. Process. (2015)
7. Srivastava, M., Panigrahi, P.K.: Quantum image representation through two-dimensional quantum states and normalized amplitude, arXiv preprint arXiv:1305.2251 (2013)
8. Venegas-Andraca, S., Ball, J.: Processing images in entangled quantum systems. Quantum Inf. Process. **9**(1), 1–11 (2010)
9. Chakraborty, S., Dey, L.: Image representation, filtering, and natural computing in a multivalued quantum system. Handbook of Research on Natural Computing for Optimization Problems, IGI-Global (2016)
10. Caraiman, S., Manta, V.: Image representation and processing using ternary quantum computing. In: Adaptive and Natural Computing Algorithms. Springer, pp. 366–375 (2013)
11. Klimov, A.B., Sánchez-Soto, L.L., de Guise, H., Björk, G.: Quantum phases of a qutrit. J. Phys. A: Math. Gen. **37**(13), 4097 (2004)
12. Zadeh, R., Haghparast, M.: A new reversible/quantum ternary comparator. Aust. J. Basic Appl. Sci. **5**(12), 2348–2355 (2011)
13. Al-Rabadi, A., Casperson, L., Perkowski, M., Song, X.: Multiplevalued quantum logic. Quantum **10**(2), 1 (2002)
14. Khan, M.H.: Design of reversible/quantum ternary multiplexer and demultiplexer. Eng. Lett. **13**(2), 65–69 (2006)
15. Muthukrishnan, A. Stroud, C. Jr.: Multivalued logic gates for quantum computation. Phys. Rev. A **62**(5) (2000)
16. Mandal, S.B., Chakrabarti, A., Sur-Kolay, S.: Quantum ternary circuit synthesis using projection operations, arXiv preprint arXiv:1205.2390 (2012)
17. Nower, N., Chowdhury, A.R.: On the realization of online ternary testable circuit. Int. J. Eng. Technol. **2**(4) (2012)

# A Graph-Theoretic Approach for Visualization of Data Set Feature Association

**Amit Kumar Das, Saptarsi Goswami, Basabi Chakraborty and Amlan Chakrabarti**

**Abstract** A graph-theoretic approach is presented in this paper to visually represent feature association in data sets. This visual representation of feature association, which has been named as Feature Association Map (FAM), is based on similarity between features measured using pair-wise Pearson's product moment correlation coefficient. Highly similar features will appear as clusters in the graph visualization. Data sets with high number of features as part of feature clusters will indicate the possibility of strong feature association. The efficacy of this method has been demonstrated in ten publicly available data sets. FAM can be applied effectively in the area of feature selection.

**Keywords** Feature association · Feature redundancy · Graph-based visualization · Feature selection

## 1 Introduction

Till the end of 1990s, very few domains were explored which included high-dimensional data sets with more than 40 features [1]. However, in the last two decades, there has been a rapid advent of biomedical research like genome projects

A.K. Das (✉) · A. Chakrabarti
A. K. Choudhury School of Information Technology,
University of Calcutta, Kolkata, India
e-mail: amitkrdas.kol@gmail.com

A. Chakrabarti
e-mail: acakcs@caluniv.ac.in

S. Goswami
Institute of Engineering and Management, Kolkata, India
e-mail: saptarsi007@gmail.com

B. Chakraborty
Iwate Prefectural University, Iwate, Japan
e-mail: basabi@iwatepu.ac.jp

© Springer Nature Singapore Pte Ltd. 2017
R. Chaki et al. (eds.), *Advanced Computing and Systems for Security*,
Advances in Intelligent Systems and Computing 568,
DOI 10.1007/978-981-10-3391-9_7

[2]. These projects have produced extremely high-dimensional data sets, ones with 20,000 or more features being very common [3]. Also, there has been a wide-spread adoption of internet and social networking leading to a need for text classification [4, 5] for customer behaviour analysis [6].

High-dimensional data sets need high amount of computational space and time. At the same time, not all features are useful [7]. Redundant features degrade the performance of models and algorithms [8, 9]. Therefore, it is quite important to understand the association between the different features to uncover feature redundancy.

One major motivation for understanding redundancy in a feature set is to be able to optimize the number of features without making much compromise on the model performance. Both in case of supervised as well as unsupervised classification activities, reducing the number of features is extremely important. This is the problem which feature selection attempts to solve [10].

Feature selection is an combinatorial optimization problem. It is ideal to exhaustively search the best subset of features amongst the $2^N$ candidate subsets, where N represents the number of features. But this exhaustive search procedure, because of the sheer size of the search space, is extremely expensive. Optimizing feature subset for a large value of N have been found as NP-complete [11]. It might reach a prohibitive level even for a mid-size value of N [7]. As an alternative to exhaustive search, approximate search technique using heuristic function in conjunction with some termination criterion can be used. In most cases, this reduces the order of the search space to O ($N^2$) or below [12].

In order to understand the association of features, especially in case of a high-dimensional data set, measuring the similarity between the features is needed. Pearson's product moment correlation coefficient is arguably one of the most common measures of feature similarity [13]. In this paper we present a novel graph-based approach, the Feature Association Map (FAM), to visually understand the association between features. FAM gives a fairly intuitive understanding of feature interaction through a simple yet powerful visualization technique. FAMs corresponding to some benchmark data sets from UCI (University of California, Irvine) repository [14] have been used to demonstrate the strength of this technique.

## 2 Related Work

Feature selection is a topic of intense research right from 1970s. However, graph theoretic approach of feature selection has started gaining research attention in the last two decades.

In one of the earliest published works in this area, an attempt has been made to leverage graph theoretic approach in feature selection [15]. This paper has presented an approach to unify a large set of algorithms in the area of both supervised and unsupervised learning into common graph theoretic framework. Using this framework, a new feature reduction technique has been proposed.

Another paper [16] has presented hyper graph-theoretic approach to feature selection based on dominant-set clustering. The feature selection approach in this paper is based on the multi-dimensional interaction information.

In another work [17], feature selection for unsupervised data has been approached in a distinct way. The entire feature set is represented as a weighted graph, then community detection algorithm is applied to it to identify feature clusters and finally an iterative search strategy based on node centrality is used for subset selection. One more approach [18] has been proposed using a dense subgraph for the unsupervised feature selection in another work.

In another recent work [19], feature selection using graph cuts based on both redundancy and relevance has been proposed as an approach especially useful for medium-size subsets.

Feature selection approach using graph theory has been used in bioinformatics too. A graph theoretical approach using cliques has been proposed resulting in significant amount of space and time savings without compromising accuracy of model performance [20].

# 3 Basic Underlying Concepts

In this section, few preliminary concepts have been outlined which is essential for the proposed approach. Grouping of vertices in a graph to form graph cluster is the first key concept. To draw the vertices, and edges between different vertices, association between features is important to understand. Association of features can be analysed using different measures. Pair-wise correlation is arguably the most commonly used measure.

## 3.1  Graph and Graph Clustering

A graph $G = (V, E, C)$ consists of a set of objects $V = \{v_1, v_2, \dots\}$ called vertices, set $E = \{e_1, e_2, \dots\}$ whose elements are called edges, such that each edge $e_k$ is identified with an unordered pair $(v_i, v_j)$ of vertices [21] and lastly set $C = \{c_1, c_2, \dots\}$ whose elements represent color of each vertex. Graphs are generally used to model different types of relations.

The goal of graph clustering is to divide a set of vertices into groups or clusters in a way such that the elements in a particular cluster are similar or connected in some way [22]. Graph clustering technique helps to detect densely connected groups of vertices or sub-graphs in a large graph [23].

## 3.2   Association Between Features

In a feature set, association between features depends on pair-wise similarity. Two features are said to be strongly associated when there is a high degree of similarity between them. Pearson's correlation coefficient is arguably one of the most common methods which indicates how similar or dissimilar the features are. This essentially means that higher the correlation between two features $f_1$ and $f_2$, more similar are the features. Hence, if the features are used for classification, the classification obtained based on feature $f_1$ will be similar to classification obtained based on feature $f_2$. This clearly indicates redundancy between two features $f_1$ and $f_2$.

There are other standard measures of feature association. For example, mutual information between features using entropy is a measure which is commonly used. However, the main advantage of using pair-wise correlation measure between features is that clear inference can be drawn from the measured value.

## 3.3   Correlation Coefficient

For a pair of variables $(X, Y)$, correlation coefficient r is defined by the equation

$$r = \frac{(cov(X, Y))}{sd(X).sd(Y)} \quad \text{where}$$

$cov(X,Y) = \sum (x_i - \bar{x}_i)(y_i - \bar{y}_i)$,
Standard Deviation $sd(X) = \sqrt{\sum (x_i - \bar{x}_i)^2}$,
Standard Deviation $sd(Y) = \sqrt{\sum (y_i - \bar{y}_i)^2}$.

Correlation coefficient has values between $-1$ and $+1$, with $+1$ indicating maximum linear correlation, $-1$ indicating minimum linear correlation and 0 indicating no correlation at all. Correlation coefficient values between 0.68–1.00 can be considered as high and values more than 0.9 can be considered as very high [24].

## 4   Proposed Approach

In the proposed approach, we attempt to represent each data set in consideration as an undirected graph or Feature Association Map (FAM). This has been done in three steps.

- Step 1: Constructing base graph
  The features of the data set will be shown as vertices of the graph. The edges between the vertices will represent the similarity between the vertices (or the features). Similarity will be measured by the pair-wise Pearson's product moment

correlation coefficient value between the features. In this approach, high pair-wise correlation value will indicate a high amount of similarity between the two features (and hence strong feature association). While drawing the edges, we will draw the ones which have a high value of pair-wise correlation only. The features that do not have any other similar feature will be represented as vertices with no connected edge (or isolated vertices). All isolated vertices will be dropped from the graph. In case of a data set having no feature similarity at all, a null graph will be generated. We will set the default color of all vertices appearing in the graph as "colour 1".

- Step 2: Identify and mark highly similar features
  We will identify the features which have very high feature similarity (or correlation) with another feature. Those are obviously features with strong association. In context of feature selection, these are the features which need to be investigated minutely for finding redundancy. In the graph we will set the color of these vertices as "colour 2".
- Step 3: Identify and highlight most critical features
  Finally, we will find out the features which have high or very high similarity with a number of features. These features will be marked as "colour 3" in the final graph. Understandably, these are the features which are most critical for consideration as they represent most number of features in terms of similarity. Hence, in context of feature selection, it is expected that these features will be a part of the optimal feature subset.

At the end of Step 3, the FAM corresponding to the data set is ready. FAM gives an immediate insight into the feature clusters for each data set. It helps in finding the hot-spots of redundancy. FAM also identifies and highlights the features which have higher similarity and sensitivity than the other features. It is particularly useful in visualization of high-dimensional data sets. This is because no other vertex except the ones in feature clusters appear in the graph. FAM is highly parametric. The ease of visualization and flexibility to experiment with different parameters is the real novelty of this approach.

We have focused on numerical variables, both continuous and discrete. Nominal and ordinal variables can be coded based on various schemes as available.

**Algorithm: Feature Association Map (FAM)**

**Input**: N-dimensional data set $D_N$, having original feature set $O = f_1, f_2, \ldots, f_N$.
**Output**: A graph $g1 = (V, E, C)$ consisting of a set of vertices, $V = v_1, v_2, \ldots, v_n$, where $n <= N$, is the size of the reduced feature set and vertices are colour-coded based on their similarity with other vertices and their criticality.

**Begin**
1: Declare N X N matrix for correlation matrix "corr-matrix" and for adjacency matrix "adj-matrix"
2: $D_{N-1} = D_N( , -N)$
3: corr-matrix $(i, j) = $ abs-correlation $(D_{N-1} (i, j))$

```
4: for each (i) do
5:    for each (j) do
6:        if (corr-matrix[i, j] ≥ α₁) then
7:            adj-matrix[i, j] = 1
8:        else
9:            adj-matrix[i, j] = 0
10:   next j
11:   adj-matrix[i, i] = 0
12: next i
13: high-conn-nodes = which(col-sum (adj-matrix) ≥ β)
14: g₁ = adjacency-graph (adj-matrix)
15: vertex-color(g₁) = colour₁
16: Set α₂ as threshold value for very high correlation
17: Repeat steps 4 to 14 with threshold = α₂ to generate graph g₂
18: for each (vertex(g₁)) do
19:   if (vertex-name(g₁)= vertex-name(g₂)) then
20:       vertex-color(g₁) = colour₂
21:       if (vertex-name(g₁)= name (high-conn-nodes)) then
22:           vertex-color(g₁) = colour₃
23: next vertex(g₁)
End
```

# 5   Experiments and Outcome

The data sets used are the benchmark data sets from the UCI Machine Learning Repository. For generating the graphs, 'igraph' library of R has been used. A summary of characteristics of the data sets used have been captured in Table 1.

Features are categorized into following sets based on similarity and interaction with other features.

**Table 1**   UCI data sets analysed

| Data set | # of features | # of instances |
|----------|---------------|----------------|
| CTG | 35 | 2126 |
| Ion | 34 | 351 |
| Mdlon | 501 | 2000 |
| Optgt | 64 | 5620 |
| Plantleaves | 64 | 4799 |
| Sonar | 60 | 208 |
| SPECTF | 44 | 267 |
| Student | 33 | 649 |
| Texture | 41 | 5500 |
| WDBC | 32 | 569 |

- Features appearing in FAM i.e., the features with high pair-wise correlation with at least one other feature. Vertices corresponding to these features are marked in "blue".
- Critical features i.e., the features with very high pair-wise correlation with at least one other feature. Vertices corresponding to these features are marked in "green".
- Extremely critical features are critical features which have high or very high pair-wise correlation with more than one feature. Vertices corresponding to these features are marked in "red".

## 5.1   Analysis of 'CTG' Data Set

Total features: 35
Number of features with high correlation: 13
Critical features: 8 ("b", "e", "LBE", "LB", "Width", "Mode", "Mean", "Median")
Extremely critical features: 6 ("LBE", "LB", "Width", "Mode", "Mean", "Median")
Graph visualization of the data set 'CTG' is given in Fig. 1.

## 5.2   Analysis of 'Ionosphere' Data Set

Total features: 34
Number of features with high correlation: 9
Critical features: 3 ("att12", "att14", "att16")

**Fig. 1**  Feature graph of 'CTG'

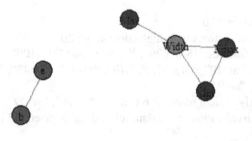

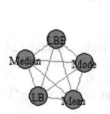

**Fig. 2**  Feature graph of
'Ionosphere'

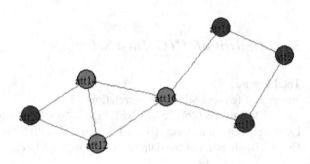

Extremely critical features: All critical features mentioned above are extremely critical

Graph visualization of the data set 'Ionosphere' is given in Fig. 2.

## 5.3  Analysis of 'mdlon' Data Set

Total features: 501
Number of features with high correlation: 20
Critical features: 18 ("at29", "at49", "at65", "at106", "at129", "at154", "at242", "at282", "at319", "at337", "at379", "at434", "at443", "at452", "at454", "at473", "at476", "at494")
Extremely critical features: 6 ("at65", "at106", "at129", "at242", "at476")
Graph visualization of the data set 'mdlon' is given in Fig. 3.

## 5.4  Analysis of 'optdgt' Data Set

Total features: 64
Number of features with high correlation: 14
Critical features: 2 ("Atr.2", "Atr.58")
Extremely critical features: 0
Graph visualization of the data set 'optdgt' is given in Fig. 4.

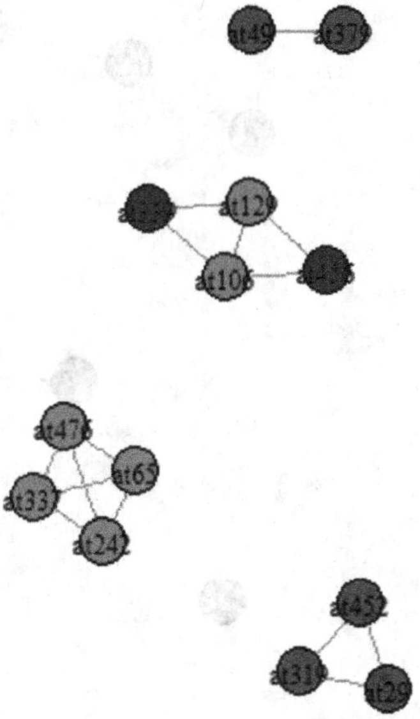

**Fig. 3** Feature graph of 'mdlon'

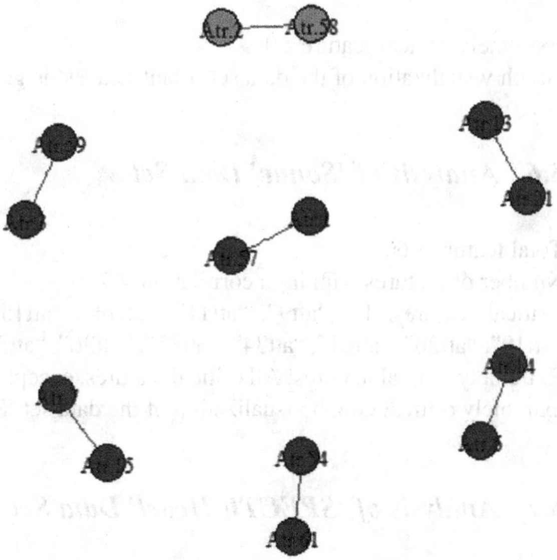

**Fig. 4** Feature graph of 'optdgt'

## 5.5 Analysis of 'Plant Leaves' Data Set

Total features: 64
Number of features with high correlation: 12
Critical features: 0

**Fig. 5** Feature graph of 'Plant Leaves'

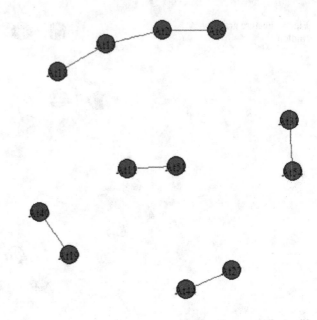

Extremely critical features: 0
Graph visualization of the data set 'Plant Leaves' is given in Fig. 5.

## 5.6   Analysis of 'Sonar' Data Set

Total features: 60
Number of features with high correlation: 49
Critical features: 17 ("att9", "att11", "att14", "att15", "att16", "att17", "att18", "att19", "att20", "att21", "att34", "att35", "att36", "att37", "att38", "att45", "att47")
Extremely critical features: All critical features, except "att20", mentioned above are extremely critical Graph visualization of the data set 'Sonar' is given in Fig. 6.

## 5.7   Analysis of 'SPECTF Heart' Data Set

Total features: 44
Number of features with high correlation: 29
Critical features: 19 ("F3S", "F4R", "F4S", "F8R", "F8S", "F9R", "F9S", "F13R", "F13S", "F15R", "F15S", "F18R", "F18S", 2F20R", "F20S", "F21R", "F21S", "F22R")
Extremely critical features: All critical features mentioned above are extremely critical
Graph visualization of the data set 'Student Performance' is given in Fig. 7.

**Fig. 6** Feature graph of
'Sonar'

**Fig. 7** Feature graph of
'SPECTF Heart'

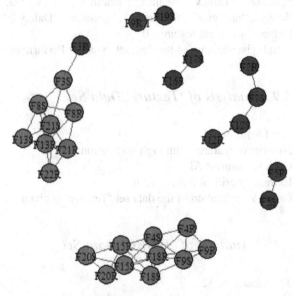

## 5.8 Analysis of 'Student Performance' Data Set

Total features: 33
Number of features with high correlation: 21
Critical features: 18 ("traveltime.x", "studytime.x", "famrel.x", "freetime.x",

**Fig. 8** Feature graph of
'Student Performance'

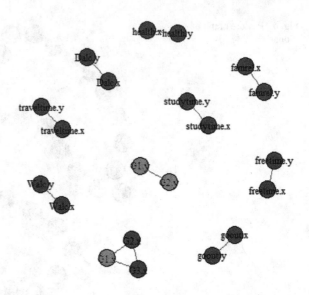

"goout.x", "Dalc.x", "Walc.x", "health.x", "G2.x", "G3.x", "traveltime.y", "study-
time.y", "famrel.y", "freetime.y", "goout.y", "Dalc.y", "Walc.y", "health.y")
Extremely critical features: 0
Graph visualization of the data set 'Student Performance' is given in Fig. 8.

## 5.9 Analysis of 'Texture' Data Set

Total features: 41
Number of features with high correlation: All
Critical features: All
Extremely critical features: All
Graph visualization of the data set 'Texture' is given in Fig. 9.

## 5.10 Analysis of 'WDBC' Data Set

Total features: 32
Number of features with high correlation: 27
Critical features: 22 ("ATT1", "ATT2", "ATT3", "ATT4", "ATT6", "ATT7", "ATT8",
"ATT11", "ATT13", "ATT14", "ATT16", "ATT17", "ATT18", "ATT20", "ATT21",
"ATT22", "ATT23", "ATT24", "ATT26", "ATT27", "ATT28", "ATT29", "ATT30")
Extremely critical features: 20 ("ATT1", "ATT3", "ATT4", "ATT6", "ATT7",
"ATT8", "ATT11", "ATT13", "ATT14", "ATT16", "ATT17", "ATT18", "ATT20",
"ATT21", "ATT23", "ATT24", "ATT26", "ATT27", "ATT28", "ATT30")
Graph visualization of the data set 'WDBC' is given in Fig. 10.

**Fig. 9** Feature graph of
'Texture'

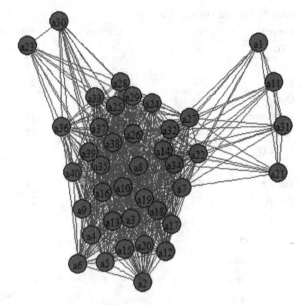

**Fig. 10** Feature graph of
'WDBC'

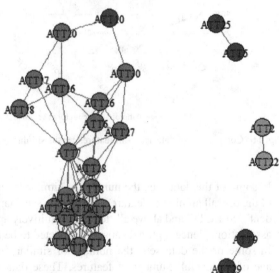

## 5.11 Summary of Outcome

Following are the salient observations that can be made from the outcome data presented in Table 2 and the visualization presented in Fig. 11.

**Table 2** Summary of outcome

| Dataset | Feats | Clusters | Simlr. feats | Critcl. feats |
|---|---|---|---|---|
| CTG | 35 | 4 | 13 | 6 |
| Ion | 34 | 2 | 9 | 3 |
| Mdlon | 501 | 7 | 20 | 6 |
| Optdgt | 64 | 7 | 14 | 0 |
| Plantleaves | 64 | 5 | 12 | 0 |
| Sonar | 60 | 2 | 49 | 16 |
| SPECTF | 44 | 6 | 29 | 19 |
| Student | 33 | 10 | 21 | 0 |
| Texture | 41 | 1 | 41 | 41 |
| WDBC | 32 | 4 | 27 | 20 |

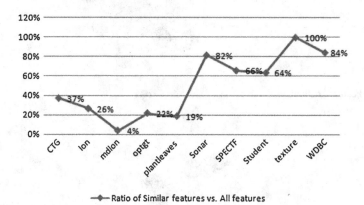

—◆— Ratio of Similar features vs. All features

**Fig. 11** Comparison of data sets—all features versus similar features

- In some of the data sets, the number of similar features is quite high percentage of the overall number of features. These data sets, namely 'Sonar', 'WDBC', 'Student', 'SPECTF' and above all 'texture', intuitively appear to have a strong feature association. Hence, optimal subset is expected to be relatively smaller in size.
- In some of the data sets, the number of similar features is quite low percentage of the overall number of features. These data sets, namely 'Ion', 'optdgt', 'plantleaves' and above all 'mdlon' intuitively appear to have a weak feature association, at least at a bivariate level.
- 'Texture' is a unique data set. All the features of this data set are highly correlated. Also, every feature has similarity with a large number of features. It seems intuitively that the optimal sub-set for 'Texture' will be quite small in size.

# 6 Conclusion

In this paper, a novel graph-based visualization technique of feature association in a data set has been presented. The current approach uses Pearson's product moment correlation coefficient as a measure of similarity between features. This gives a firm basis for the visualization technique. This novel but simple approach can be used either independently or in conjunction with other feature selection approaches to understand the basic nature of data set features.

This work can be extended to capture graphs based on the other similarity measures e.g., mutual information and Fisher score. Also, different principles of graph theory can be applied to come up with novel ways to identify optimal feature subsets in different situations and with different types of data set. These subsets can be evaluated using different models to draw useful inferences.

# References

1. Elisseeff, A., Guyon, I.: An introduction to variable and feature selection. J. Mach. Learn. Res. **3**, 1157–1182 (2003)
2. Xing, E., Jordan, M., Karp, R.: Feature selection for high-dimensional genomic microarray data. In: Proceedings of the Eighteenth International Conference on Machine Learning, pp. 601–608 (2001)
3. Dougherty, E.R., Hua, J., Tembe, W.: Performance of feature-selection methods in the classification of high-dimension data. Pattern Recognit **42**, 409–424 (2009)
4. Yang, Y., Pederson, J.O.: A comparative study on feature selection in text categorization. In: Proceedings of the Fourteenth International Conference on Machine Learning, pp. 412–420 (1997)
5. Dey Sarkar, S., Goswami, S., Agarwal, A., Aktar, J.: A novel feature selection technique for text classification using naive bayes. Int. Sch. Res. Notices (2014)
6. Ng, K., Liu, H.: Customer retention via data mining. AI Rev, **14**, 569–590 (2000)
7. Dash, M., Liu, H.: Feature selection for classifications. Intell. Data Anal. Int. J. **1**, 131–156 (1997)
8. Liu, H., Yu, L.: Feature selection for high-dimensional data: a fast correlation-based filter solution. In: International Conference on Machine Learning (2003)
9. Goswami, S., Chakrabarti, A.: Feature selection: a practitioner view. Int. J. Inf. Technol. Comput. Sci. (IJITCS) **6**(11), 66 (2014)
10. Duda, R., Hart, P., Stork, D.G.: Pattern Classification, 2nd edn. Wiley, New York (2001)
11. Blum, A.L., Rivest, R.L.: Training a 3-Node Neural Network is NP-Complete. In: COLT (1988)
12. John, G.H., Kohavi, R., Pfleger, K.: Irrelevant Features and the Subset Selection Problem. In: ICML (1994)
13. John, G.H., Kohavi, R.: Wrappers for feature subset selection. Artif. Intell. **97**, 273–324 (1997)
14. Bache, K., Lichman, M.: UCI machine learning repository. University of California, Irvine, School of Information and Computer Sciences (2013)
15. Lin, S., Xu, D., Yan, S., Yang, Q., Zhang, B., Zhang, H.: Graph embedding and extensions: a general framework for dimensionality reduction. IEEE Trans. Pattern Anal. Mach. Intell. **29**, 40–51 (2007)
16. Hancock, E.R., Zhang, Z.: A Graph-Based Approach to Feature Selection. In: GBRPR (2011)
17. Moradi, P., Rostami, M.: A graph theoretic approach for unsupervised feature selection. Eng. Appl. AI **44**, 33–45 (2015)

18. Bandyopadhyay, S., Bhadra, T., Mitra, P., Maulik, U.: Integration of dense subgraph finding with feature clustering for unsupervised feature selection. Pattern Recognit. Lett. **40**, 104–112 (2014)
19. Ishii, M., Sato, A.: Feature selection using graph cuts based on relevance and redundancy. In: ICIP (2013)
20. Altun, G., Gremalschi, S., Hu, H., Harrison, R.W., Pan, Y.: A feature selection algorithm based on graph theory and random forests for protein secondary structure prediction. In: ISBRA (2007)
21. Deo, N.: Graph Theory with Applications to Engineering and Computer Science, Eastern Economy Edition (1974)
22. Schaeffer, S.E.: Graph clustering. Comput. Sci. Rev. **1**, 27–64 (2007)
23. Cheng, H., Yu, J.X., Zhou, Y.: Graph clustering based on structural/attribute similarities. PVLDB **2**, 718–729 (2009)
24. Taylor, R.: Interpretation of the correlation coefficient: a basic review. J. Diagn. Med. Sonography, **6**(1), 35–39 (1990)

# A Novel Approach for Human Silhouette Extraction from Video Data

**Amlan Raychaudhuri, Satyabrata Maity, Amlan Chakrabarti and Debotosh Bhattacharjee**

**Abstract** In this paper we propose a method for efficient extraction of human silhouette from video sequences. The proposed approach includes background elimination, edge detection, region filling and noise removal using morphological operations to estimate the silhouette of an image. To the best of our knowledge our proposed approach for silhouette extraction involving background elimination and edge detection is first of its kind. We have applied our proposed technique on Weizmann (standard) dataset and compared the results with the most recent related research work. The comparison results in terms of statistical measures like precision, recall and F-measure clearly show the supremacy of our method and thus justify its novelty.

**Keywords** Human silhouette · Foreground detection · Edge detection · Background subtraction · Morphological operation · Region filling

A. Raychaudhuri (✉)
Department of Computer Science & Engineering, B. P. Poddar Institute
of Management & Technology, 137, VIP Road, Kolkata-52, India
e-mail: amlanrc@gmail.com

S. Maity · A. Chakrabarti
A. K. Choudhury School of Information Technology, University of Calcutta,
JD-2, JD Block, Sector-III, Kolkata-98, India
e-mail: satyabrata.maity@gmail.com

A. Chakrabarti
e-mail: amlanc@ieee.org

D. Bhattacharjee
Department of Computer Science and Engineering, Jadavpur University,
188, Raja S. C. Mallick Road, Kolkata-32, India
e-mail: debotoshb@hotmail.com

© Springer Nature Singapore Pte Ltd. 2017 125
R. Chaki et al. (eds.), *Advanced Computing and Systems for Security*,
Advances in Intelligent Systems and Computing 568,
DOI 10.1007/978-981-10-3391-9_8

# 1   Introduction

Human silhouette extraction from video is an important step towards shape based analysis for many human based video applications. Some of the video applications like CCTV camera for indoor surveillance, elderly monitoring, and office floor surveillance etc., where camera is fixed in certain position, extraction of background information is an important step. Our proposed methodology eliminates the background efficiently to accurately extract human silhouettes as foreground objects. Sometimes the color of the foreground is closely matched with some portion of the background, that results distorted foreground detection. Silhouette extraction is an important step for segmenting a human body from a background, which can be used to track the person in video, recognizing human actions or gait based biometric recognition.

In the process of human silhouette extraction, background modeling is one of the important tasks. Background subtraction has been an active area of research, and it is a commonly used approach for detecting as well as tracking a person in videos from a fixed camera. An efficient thresholding mechanism is needed after eliminating the background from the current frame to get the noise free foreground. Background subtraction methods have been used extensively to assist in human behavior analysis such as human action recognition and human gait recognition [1–7], where accurate human silhouettes are needed to extract the features of human body configuration. However, the human silhouettes acquired by background subtraction are usually not accurate enough for the recognition tasks; in particular, shadows are not properly removed from human silhouettes. Also, the cases in which foreground object passes the background areas with almost identical intensity values to those of the human body, background subtraction methods cannot give accurate human silhouettes. These are the problems which decrease the recognition performances significantly [8]. Horprasert et al. [9] proposed a robust background subtraction and shadow detection technique, which was applied to an object tracking system alongside with appearance model [10]. Ahn et al. [11] proposed a method for human silhouette extraction which is based on background subtraction for different regions, where regions are considered as visual primitives, rather than pixels.

In [12], Schreer et al. proposed a method to extract silhouettes in a YUV color space, which is able to detect shadow and eliminate it in real time. But this algorithm does not give good result when there exists a static target object or when there are other moving objects in the scene.

In [13, 14], the authors have extracted the human silhouette from video sequence. They have identified the moving object from the video sequence by background subtraction which is followed by several steps for extracting the silhouette.

In [15], shadow evaluator is used to verify each raw shadow pixel which was detected through Gaussian distribution analysis. The authors also proposed a silhouette compensation technique to recover some missing silhouette pixels.

In this research work we perform background subtraction on the current frame of a video as the first step to detect the foreground. The human silhouette, which is the intended foreground object, is extracted based on a threshold. There are chances that some of the human silhouette pixels may be wrongly treated as background pixel. Those pixels are recovered by region filling method and finally noise is removed from the outside of human silhouette boundary by using morphological operation. The remainder of this paper is organized as follow: Sect. 2 yields details of the proposed method. Experimental results and discussion are described in Sect. 3. Finally Sect. 4 concludes the paper.

## 2 Proposed Method

We perform extraction of frames from the input video and convert them to gray level frames. Next, a new image is created by subtracting the background frame from the current frame. Foreground is detected from the subtracted image by using a threshold value. Boundary edge of a person is also detected from the subtracted image and then we merge the foreground and the edge detected images. Standard morphological operations are applied on the merged image to generate the final silhouette. The brief illustration of the proposed methodology is given in Fig. 1.

The detailed description of the proposed technique is given below.

### 2.1 Foreground Detection

Foreground detection is done based on the rule described in Eq. (1).

$$I1\,(x,y) = \begin{cases} 1, & if \ |I\,(x,y) - B(x,y)| \rangle \sigma \\ 0, & otherwise \end{cases} \tag{1}$$

Where $I\,(x, y)$ is the intensity value of the current frame for the pixel coordinate $(x, y)$ and $B\,(x, y)$ is the intensity value of the background frame for the same pixel. $I1\,(x, y)$ will be treated as foreground pixel if the absolute value of the background subtraction from the current frame is greater than the threshold value $\sigma$, which is the standard deviation of the current frame. Otherwise it will be treated as background pixel. The value of $\sigma$ is calculated using Eqs. (2) and (3). Here standard deviation is used as a threshold value because it quantifies the amount of variation of pixel intensity from the mean intensity of an image.

$$\sigma = \sqrt{\left( \sum_1^r \sum_1^c (I(x,y) - \mu)^2 \right) / n} \tag{2}$$

$$\mu = \left( \sum_1^r \sum_1^c I(x,y) \right) / n \tag{3}$$

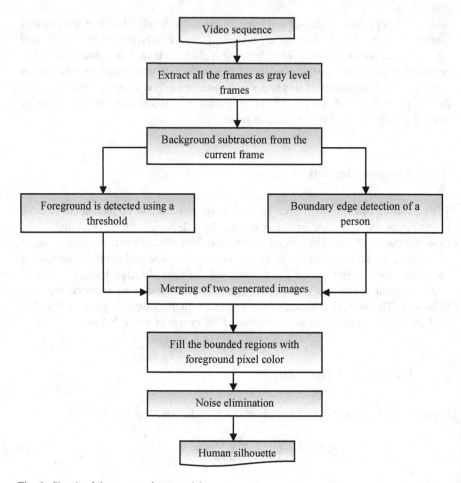

**Fig. 1** Sketch of the proposed approach

In Eqs. (2) and (3), $r$ and $c$ denote the number of pixels row wise and the number of pixels column wise of the current frame respectively. The value of $n$ represents the total number of pixels of the current frame and $\mu$ represents the mean intensity value of the current frame. After foreground detection, $I1$ image is created.

## 2.2 Boundary Edge Detection of a Human

After subtraction of background frame from the current frame, a difference image is generated. Then we have applied Sobel operator [16] for edge detection. Sobel operator is used here because; it gives an estimate of edge direction and edge magnitude at a point and thus generating good edge information. As a result, a new image $I2$ is created in which a human's edge boundary is detected for the current frame.

## 2.3 Merging of Two Generated Images

After creation of two images $I1$ and $I2$ using the above mentioned techniques, they are merged together using Eq. (4), and as a result a new image $I3$ is generated. For the $I3$ image, $(x, y)$ pixel coordinate will be treated as foreground pixel if $(x, y)$ pixel of $I1$ image is foreground pixel or $(x, y)$ pixel of $I2$ image is foreground pixel. Otherwise $I3(x, y)$ will be treated as a background pixel.

$$I3(x,y) = \begin{cases} 1, & if \, I1(x,y) = 1 \, or \, I2(x,y) = 1 \\ 0, & otherwise \end{cases} \tag{4}$$

## 2.4 Morphological Operations

Morphological operations are used to reduce the effect of unwanted noise after combining two different information, as done in the previous steps to generate the final silhouette image. In the $I3$ image, some human silhouette pixels are wrongly detected as background pixels and some background pixels are also wrongly treated as foreground pixels. Those misclassifications can be reduced by applying two morphological operations consecutively on the image $I3$. At the first step, fill the bounded regions with foreground color which is surrounded by edge boundary. This step is necessary if some foreground pixels are treated as background pixels within the foreground boundary. There may be some noises outside the silhouette region in the resultant image. In the next step, noises are removed from outside the human silhouette with the help of the morphological *open* operation.

We have applied our proposed technique on video sequences of Weizmann dataset [17]. The output after each step of our proposed approach for a sample current frame is shown in Fig. 2. The sample current frame is shown in Fig. 2a. Figure 2b shows the output of the foreground detection technique which is achieved by using a threshold on the image, which is generated by background subtraction from the current frame as discussed in Sect. 2.1. The resultant image contains some deformities in edge area and in some inner portion. At the next step, Sobel edge detection technique is performed on the subtracted image to get the boundary edges of a human, which resolves the edge deformation. This step has been described in Sect. 2.2 and the corresponding output is shown in Fig. 2c. Next a new image is created by merging the two images which we have generated in the previous two steps. The detailed of this step has been described in Sect. 2.3 and the corresponding output is shown in Fig. 2d. Then a morphological operation is performed on the merged image to fill the bounded edge region with the foreground color to resolve the deformities in the inner portion of the silhouette, which is described in Sect. 2.4 and the corresponding output is shown in Fig. 2e. In the next step, noises outside the human silhouette region is removed using the morphological *open* operation and thus the final silhouette is generated. This step is described in Sect. 2.4 and the corresponding output is shown in Fig. 2f.

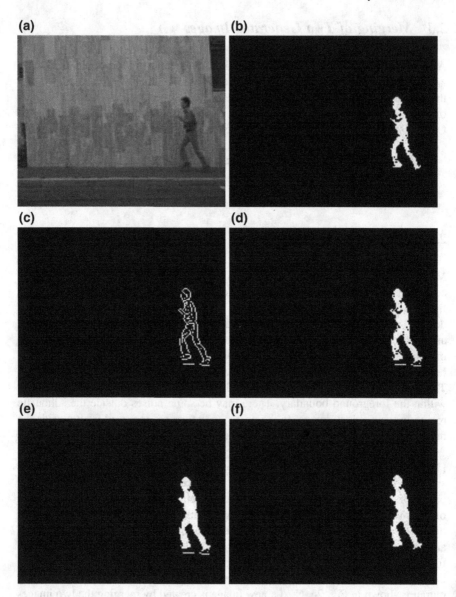

**Fig. 2** A sample input frame and the output of the corresponding frame after each step of our proposed method. **a** Sample current frame. **b** Foreground detection. **c** Edge detection. **d** Merging of previous two images. **e** Fill the bounded regions. **f** Final silhouette

# 3   Experimental Results and Discussion

To evaluate the effectiveness of our approach, we performed experiments on several video sequences of Weizmann dataset [17]. The Weizmann data is a popular single-view action data set which contains video sequences for ten different actions performed by nine actors. It provides 180 × 144 resolution images for all the videos. For effectiveness of our work, we compared it with the readily available silhouettes in Weizmann dataset, which are treated as ground truth. For different actions' sample current frames, their corresponding ground truth silhouettes and silhouettes of our proposed approach are shown in Fig. 3.

**Fig. 3** **a** Sample current frame of different actions **b** Corresponding Ground Truth Silhouette **c** Corresponding Silhouette by Proposed approach

To evaluate the performance of the proposed method, three metrics *Precision, Recall* and *F-measure* [18] have been used in this paper. These metrics are calculated with the Eqs. (5)–(7).

$$Precision = TP/(TP + FP) \tag{5}$$

$$Recall = TP/(TP + FN) \tag{6}$$

$$F - measure = 2*Precision*Recall/(Precision + Recall) \tag{7}$$

*True positives (TP)* are positive examples correctly labeled as positives. *False positives (FP)* mean negative examples incorrectly labeled as positive and *false negatives (FN)* refer to positive examples incorrectly labeled as negative. The *average Precision (AP)* and *average Recall (AR)* results of our proposed approach for seven different actions (*side, jump, jack, pjump, bend, wave1* and *wave2*) are given in Table 1. Table 2 shows the *average Precision (AP)* and *average Recall (AR)* results of our proposed approach for three other different actions (*walk, run* and *skip*). All the ten actions are performed by nine actors. In the video dataset for *walk, run and skip* actions, there are two separate video sequences of *lena (lena1 and lena2)* for each of the actions. Whereas for all other actions we have the single video sequence of *lena*. It is shown from the tables that the *average Precision (AP)* and the *average Recall (AR)* are high for all the video sequences.

Our proposed approach is compared with Shaikh et al. [13] approach based on average F-measure value for each of the ten actions of Weizmann data, which is shown in Table 3. The comparison is also shown using graph in Fig. 4. From Table 3 as well as from Fig. 4, it is clearly shown that our proposed method provides better average *F-measure* values compared to the other method for all the ten actions. Thus, it establishes the effectiveness of our proposed approach.

**Table 1** Average Precision (AP) and Average Recall (AR) values for *side, jump, jack, pjump, bend, wave1* and *wave2* actions

| Videos | Side | | Jump | | Jack | | Pjump | | Bend | | Wave1 | | Wave2 | |
|--------|------|------|------|------|------|------|------|------|------|------|------|------|------|------|
| | AP | AR | AP | AR | AP | AR | AP | AR | AP | AR | AP | AR | AP | AR |
| Daria | 0.99 | 0.89 | 0.99 | 0.90 | 0.88 | 0.96 | 0.89 | 0.95 | 0.91 | 0.92 | 0.91 | 0.97 | 0.90 | 0.97 |
| Denis | 0.97 | 0.85 | 0.97 | 0.88 | 0.81 | 0.87 | 0.89 | 0.89 | 0.87 | 0.82 | 0.88 | 0.88 | 0.86 | 0.88 |
| Eli | 0.96 | 0.96 | 0.98 | 0.95 | 0.83 | 0.98 | 0.85 | 0.99 | 0.82 | 0.98 | 0.82 | 0.97 | 0.76 | 0.96 |
| Ido | 0.97 | 0.94 | 0.98 | 0.96 | 0.93 | 0.86 | 0.93 | 0.96 | 0.91 | 0.91 | 0.96 | 0.92 | 0.94 | 0.93 |
| Ira | 0.97 | 0.80 | 0.98 | 0.78 | 0.85 | 0.80 | 0.89 | 0.89 | 0.88 | 0.86 | 0.89 | 0.91 | 0.88 | 0.88 |
| Lena | 0.99 | 0.87 | 0.99 | 0.90 | 0.91 | 0.97 | 0.94 | 0.94 | 0.90 | 0.96 | 0.91 | 0.98 | 0.90 | 0.98 |
| Lyova | 0.97 | 0.93 | 0.98 | 0.94 | 0.87 | 0.87 | 0.86 | 0.94 | 0.87 | 0.88 | 0.89 | 0.88 | 0.89 | 0.92 |
| Moshe | 0.99 | 0.87 | 0.99 | 0.90 | 0.88 | 0.97 | 0.89 | 0.97 | 0.89 | 0.98 | 0.90 | 0.99 | 0.88 | 0.99 |
| Shahar | 0.83 | 0.81 | 0.88 | 0.77 | 0.89 | 0.97 | 0.86 | 0.75 | 0.83 | 0.84 | 0.85 | 0.76 | 0.83 | 0.76 |

**Table 2** Average Precision (AP) and Average Recall (AR) values for *walk, run* and *skip* actions

| Videos | Walk | | Run | | Skip | |
|---|---|---|---|---|---|---|
| | AP | AR | AP | AR | AP | AR |
| Daria | 0.99 | 0.88 | 0.99 | 0.85 | 0.99 | 0.90 |
| Denis | 0.98 | 0.81 | 0.98 | 0.78 | 0.98 | 0.80 |
| Eli | 0.98 | 0.86 | 0.99 | 0.87 | 0.99 | 0.88 |
| Ido | 0.99 | 0.92 | 0.98 | 0.92 | 0.98 | 0.92 |
| Ira | 0.99 | 0.75 | 0.99 | 0.76 | 0.99 | 0.76 |
| Lena1 | 0.99 | 0.90 | 0.99 | 0.89 | 0.99 | 0.91 |
| Lena2 | 0.99 | 0.89 | 0.99 | 0.90 | 0.99 | 0.90 |
| Lyova | 0.99 | 0.91 | 0.98 | 0.91 | 0.98 | 0.93 |
| Moshe | 0.99 | 0.87 | 0.99 | 0.86 | 0.98 | 0.87 |
| Shahar | 0.98 | 0.91 | 0.99 | 0.89 | 0.99 | 0.89 |

**Table 3** Average F-measure based comparison result for each action

| Methods | Side | Jump | Jack | Pjump | Bend | Wave1 | Wave2 | Walk | Run | Skip |
|---|---|---|---|---|---|---|---|---|---|---|
| Shaikh et al. method [13] | 0.839 | 0.804 | 0.847 | 0.876 | 0.854 | 0.868 | 0.849 | 0.835 | 0.821 | 0.841 |
| Proposed method | **0.917** | **0.926** | **0.893** | **0.903** | **0.889** | **0.902** | **0.893** | **0.924** | **0.92** | **0.927** |

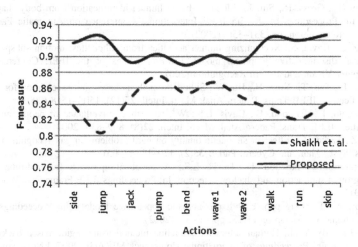

**Fig. 4** Comparison based on F-measure

# 4 Conclusion

This paper proposes an approach for extracting efficient human silhouette from video sequences. Nowadays, accurate human silhouette generation from a video sequence is very useful in various application areas like, gait recognition, human action recognition, human detection and tracking from videos, etc. The proposed method can also be applied to get the silhouettes for other moving objects (animal, car, etc.), which will be useful for object recognition and classification related problem. The proposed algorithm showed excellent results for the above mentioned video sequences. The limitation of this method is that the camera should be static throughout the video sequence. In future, we wish to extend this work to find an efficient human silhouette extraction procedure which can work when a human's body color or cloth color is very close to that of the background color.

# References

1. Liu, Z., Sarkar, S.: Improved gait recognition by gait dynamics normalization. IEEE Trans. Pattern Anal. Mach. Intell. **28**(6), 863–876 (2006)
2. Li, H., Greenspan, M.: Multi-scale gesture recognition from time-varying contours. In: IEEE International Conference Computer Vision, pp. 236–243, (2005)
3. Dedeoglu, Y., Toreyin, B., Gudukbay, U., Cetin, A.: Silhouette based method for object classification and human action recognition in video. In: Computer Vision in Human-Computer Interaction, LNCS, vol. 3979, pp. 64–77, (2006)
4. Collins, R., Gross, R., Shi, J.: Silhouette-based human identification from body shape and gait. In: Proceedings of the 5th IEEE International Conference on Automatic Face and Gesture Recognition, pp. 351–356, (2002)
5. Wang, L., David, S.: Recognizing human activities from silhouettes: motion subspace and factorial discriminative graphical model. In: Proceedings of the IEEE Conference on Computer Vision and Pattern Recognition (2007)
6. Wang, L., Tan, T., Ning, H., Hu, W.: Silhouette analysis based gait recognition for human identification. IEEE Trans. Pattern Anal. Mach. Intell. **25**(12), 1505–1518 (2003)
7. Haritaoglu, I., Harwood, D., Davis, L.S.: W$^4$: Real time surveillance of people and their activities. IEEE Trans. Pattern Anal. Mach. Intell. **22**(8), 809–830 (2000)
8. Liu, Z., Sarkar, S.: Effect of silhouette quality on hard problems in gait recognition. IEEE Trans. Syst., Man, and Cybern. Part B **35**(2), 170–183 (2005)
9. Horprasert, T., Harwood, D., Davis, L.S.: A statistical approach for real-time robust background subtraction and shadow detection. In: Proceedings IEEE Frame Rate Workshop, (1999)
10. Senior, A.: Tracking people with probabilistic appearance models. In: Proceedings IEEE International Workshop on PETS, pp. 48–55, (2002)
11. Ahn, J.H., Byun, H.: Human silhouette extraction method using region based background subtraction. In: Proceedings of International Conference MIRAGE 2007, LNCS, vol. 4418, pp. 412–420, (2007)
12. Schreer, O., Feldmann, I., Golz, U., Kauff, P.: Fast and robust shadow detection in videoconference applications. In: Proceedings of 4th EURASIP-IEEE Region 8 International Symposium on VIPromCom, pp. 371–375, (2002)

13. Shaikh, S.H., Bhunia, S.K., Chaki, N.: On generation of silhouette of moving objects from video. In: Proceedings of 4th International Conference on Signal and Image Processing, LNEE, vol. 221, pp. 213–223, (2013)
14. Shaikh, S.H., Saeed, K., Chaki, N.: Moving Object Detection Using Background Subtraction. In: Springer Briefs in Computer Science, pp. 25–48, (2014)
15. Wang, Z., Shin, B.S., Klette, R.: Accurate silhouette extraction of a person in video data by shadow evaluation. Int. J. Comput. Theor. Eng. $6$(6), 476–483 (2014)
16. Gonzalez, R.C., Woods, R.E.: Digital Image Processing, 2nd edn. Pearson Education, (2002)
17. Blank, M., Gorelick, L., Shechtman, E., Irani, M., Basri, R.: Actions as space-time shapes. In: Proceedings of IEEE International Confernces on Computer Vision, (2005)
18. Intawong, K., Scuturici, M., Miguet, S.: A New pixel-based quality measure for segmentation algorithms integrating precision, recall and specificity. In: Proceedings of CAIP 2013, Part I, LNCS, vol. 8047, pp. 188–195, (2013)

# Part IV
# Pattern Recognition

# Statistical Textural Features for Text-Line Level Handwritten *Indic* Script Identification

Pawan Kumar Singh, Ram Sarkar and Mita Nasipuri

**Abstract** As India is a multilingual country, hence, a variety of scripts are used here to write different languages. However, it becomes essential to recognize a particular script before the selection of an appropriate Optical Character Recognition (OCR) system. The research in this field is comparatively less explored and further research is required, particularly in the field of handwritten documents. This paper presents a robust script identification technique for 11 official handwritten *Indic* scripts *namely, Bangla, Devanagari, Gujarati, Gurumukhi, Kannada, Malayalam, Manipuri, Oriya, Tamil, Telugu, Urdu* along with *Roman* script. The recognition is performed at text-line level by using statistical textural features called Neighborhood Gray-Tone Difference Matrix along with Gray-level Run Length Matrix. The proposed method is experimented on a total dataset of 2400 handwritten text-lines of various scripts and yielded an identification rate of 97.69% using Multi Layer Perceptron (MLP) classifier.

**Keywords** Handwritten script identification · *Indic* scripts · Statistical textural features · Neighborhood gray-tone difference matrix · Gray-level run length matrix · Multiple classifiers

P.K. Singh (✉) · R. Sarkar · M. Nasipuri
Department of Computer Science and Engineering,
Jadavpur University, Kolkata 700032, West Bengal, India
e-mail: pawansingh.ju@gmail.com

R. Sarkar
e-mail: raamsarkar@gmail.com

M. Nasipuri
e-mail: mitanasipuri@gmail.com

© Springer Nature Singapore Pte Ltd. 2017
R. Chaki et al. (eds.), *Advanced Computing and Systems for Security*,
Advances in Intelligent Systems and Computing 568,
DOI 10.1007/978-981-10-3391-9_9

# 1   Introduction

Script is considered as a graphic form which is used in any writing system. The languages used in the human society are typeset with the different scripts. A script can be used by only one language or it can be shared by several languages, with or without any variations [1]. India has 23 languages [2], recognized by constitution, *viz., Assamese, Bengali, Bodo, Dogari, Kannada, Hindi, Sindhi, Nepali, Urdu, Punjabi, Marathi, Gujarati, Oriya, Sanskrit, Tamil, Telugu, Malayalam, Kashmiri, Manipuri, Konkani, Maithali, Santhali* and *English.* The 12 major scripts used to write these languages are: *Bangla, Devanagari, Gujarati, Gurumukhi, Manipuri, Malayalam, Oriya, Tamil, Telugu, Kannada, Roman* and *Urdu.* Among these, only *Urdu* is written from right to left whereas the rest of the scripts are written from left to right. The first 10 scripts, originated from the ancient *Brahmi* script, are also known as *Indic* scripts.

In general, any OCR system is used to recognize only a script of particular type, and for the same reason, it is not viable to model a single OCR system for recognizing variety of scripts/languages. Hence, one can think of making a pool of OCR engines which correspond to different scripts in a multi-lingual environment. However, for this, it is necessary to have the knowledge about the script used to write the document. Hence, this necessity could be fulfilled if the researchers can be able to design an automatic script recognition module for the multi-script scenario.

An automatic script identification module would be used to sort or search the relevant information when the domain is multilingual/multi-script or even it helps to index/categorize the documents images on the basis of its script type. When a script is used to write only one language, then the script recognition technique can also be considered as language recognition technique. Otherwise, script recognition is the first step of classification followed by language identification among the languages which share a common script.

Different methodologies have been reported in the literature for accomplishing this task, sometimes with high degree of accuracy. A comprehensive survey based on script recognition techniques had been prepared by Singh et al. [1], with emphasis on script identification in both printed and handwritten *Indic* scripts scenario. Script identification for printed documents at page level [3–5], text-line level [6–10], and word level [11–16] have been found in the literature. On the contrary, only few works have been done for handwritten script identification at page level [17], text-line level [18, 19], and word level [20–22]. Singh et al. [17] developed a page-level script identification technique for handwritten document pages using Gray Level Co-occurrence Matrix (GLCM). The proposed technique had been experimented on four scripts *namely, Bangla, Devanagari, Telugu,* and *Roman* and the system was found to identify 91.48% scripts successfully using MLP classifier. Hangarge et al. [18] described a set of 13 spatial spread features of the three scripts *namely*, *English*, *Devanagari* and *Urdu* which were extracted using morphological filters. Experiments were carried out with *k*-NN classifier by varying the number of neighbors ($k$ = 3, 5, 7, 9, 11, 13, 15) and the performance of the technique was optimal when the value of $k$ is set to 3. The overall recognition accuracies of the proposed system were found to be 88.67% and 99.2% for tri-script and bi-script cases respectively. Singh et al. [19] proposed a texture based concept for script identification at text-line level for six handwritten scripts *namely*, *Bangla*, *Devanagari*, *Malayalam*, *Tamil*, *Telugu* and *Roman*. An accuracy of 95.67% had been achieved using 3-fold cross validation of MLP classifier. Roy et al. [20] described a scheme for word-wise identification of handwritten *Roman* and *Oriya* scripts for Indian postal automation using water-reservoir and topological features. The overall accuracy rate achieved on the test dataset was found to be 99.6% and 97.69% respectively. Sarkar et al. [21] presented a system, which identified the scripts of the handwritten words from the document images, written in *Bangla* or *Devanagari* mixed with *Roman* scripts with the help of eight holistic features. The recognition performances of 99.29% and 98.43% had been achieved on the test sets of *Bangla-English* words and *Devanagari-English* words respectively. P.K. Singh et al. reported a technique [22] which recognized the scripts of handwritten words from a document page, written in *Devanagari* script mixed with *Roman* script. A set of 39 distinctive features using topological along with convex hull based features were designed for the recognition purpose and the overall script identification accuracy of 99.54% was achieved. However, a major limitation of the above works is that researchers have considered only a few *Indic* scripts. This has been a major point of motivation behind developing a robust handwritten script identification technique including all the official *Indic* scripts along with *Roman* script.

## 2 Challenges Related to Handwritten Script Identification

There are some unique challenges that must be addressed in the domain of hand-written script recognition system. Among many, two basic problems are: *inter-writer* variability and *intra-writer* variability. *Inter-writer* variability encompasses the variations seen among different writers i.e., different writers will invariably have different writing styles. In contrast, *intra-writer* variability takes into consideration that the same writer tends to write the same textual content in a different manner depending upon his/her frame of mind. The challenge in this regard is to create a writer-independent script identification system that has the ability to adapt these variations like humans. Another major challenge that the system has to address is the problem of constrained versus unconstrained handwriting. Constrained hand-writing refers to handwritten text that conforms to a pre-defined writing constriction, e.g. all the text words in a document image will be discrete and non-touching.

Whereas unconstrained handwriting refers to the fact that the document image may contain discrete and cursive handwriting or a mixture of both, with no restriction on the writers while they write. Apart from this, difficulties inherent in recognizing handwritten scripts pose huge challenges than its printed form. Similarity among different scripts is quite common when the documents are handwritten. The styles of writing for handwritten scripts are more diverse than printed fonts. Also, problems such as existence of ruling lines, noise, skew, quality of ink, age of the document, etc. are commonly seen in handwritten documents. As mentioned earlier, script identification can be achieved at either page level, text-line level or word level. Sometimes, identifying scripts at page-level can be sometimes too convoluted and protracted due to large computational complexity. On the other hand, identifying words written in different scripts using a few characters is definitely a challenging task because the number of characters presents in a single word may not produce significant amount of discriminative information required for identification. Therefore, considering the complexities of the scripts, it would be better to identify the scripts at text-line level compared to page or word-level.

## 3   Data Collection and Pre-processing

There are no standard databases, considering either handwritten or printed *Indic* script documents, available in public domain which can be successfully used for this experimentation. Hence, we have prepared in-house database of handwritten documents. Different educated people were requested to write few text-lines of his/her choice inside A-4 size pages. Handwritten text-lines were written in 12 official scripts of India as mentioned earlier. It is to be noted that writers involved in the data collection drive belong to different professions. These pages are then scanned at 300 dpi resolution and saved as gray tone images. The noisy pixels therein, if any, are removed by Gaussian filter [23]. It is worth mentioning that the *inter*-word and *intra*-word spaces are very non-uniform in the handwritten text-lines. Numerals of any script which may be present in the text document are not considered for the present work. A text–line whose width is at least 50% of the page width is considered here. A sample snapshot of text-line images written in 12 different scripts is shown in Fig. 1. Otsu's global thresholding approach [24] is used to convert them into two-tone images (0 and 1) where the label '1' represents the object and '0' represents the background. However, the dots and punctuation marks appearing in the text- lines have not been eliminated, since these may also contribute to understand the text in a meaningful way. Finally, 2400 handwritten text-line images are prepared with precisely 200 text-lines per script.

**Fig. 1** Sample text-line images written in: **a** *Bangla,* **b** *Devanagari,* **c** *Gujarati,* **d** *Gurumukhi,* **e** *Kannada,* **f** *Malayalam,* **g** *Manipuri,* **h** *Oriya,* **i** *Tamil,* **j** *Telugu,* **k** *Urdu,* and **l** *Roman* scripts respectively

## 4 Proposed Work

Every script/language, consisting of a finite set of characters, has a distinct visual appearance, which serves as useful visual clues to recognize the script. The current research is inspired by this simple observation of the human beings which also motivates the researchers to design different texture based features. Usually, texture features are designed to capture the granularity and repetitive patterns of local regions seen within an image. Some well-known texture features relying on GLCM and Gabor filter bank consider multiple scales and orientations for feature extraction which in turns involves a high computation cost. The conventional statistical textural features utilized in this paper, are Neighborhood Gray-Tone Difference Matrix (NGTDM) and Gray-level Run Length Matrix (GLRLM). These features are illustrated below in the following subsections.

## 4.1   Neighborhood Gray-Tone Difference Matrix (NGTDM)

A NGTDM [25] defines the texture measures which are very much correlated with human perception of textures. It calculates the texture using neighborhood intensity differences which will be helpful to describe the local features. The NGTDM are based on the differences between each pixel and the neighboring pixels in the adjacent regions. A NGTDM is basically a column vector of $G$ elements. This vector is populated by computing the difference between the intensity values of a pixel and the mean intensity calculated over a square shaped window centered at that pixel. Suppose the image intensity level $f(x, y)$ at location $(x, y)$ is $i$, $i = 0, 1, 2, \ldots, L - 1$. The mean intensity value of the window centered at $(x, y)$ can be written as:

$$f_i = f(x, y) = \frac{1}{W - 1} \sum_{m = -K}^{K} \sum_{n = -K}^{K} f(x + m, y + n) \tag{1}$$

where, $K$ denotes the window size and $W = (2K + 1)^2$. The $i$-th entry of the gray-tone difference matrix is given by:

$$g(i) = \sum_{x=0}^{M-1} \sum_{y=0}^{N-1} |i - f_i| \tag{2}$$

for the pixels whose intensity value is $i$. Otherwise, $g(i) = 0$.

Five different features are derived from the NGTDM, described below, to quantitatively estimate the following perceptual texture properties:

- **Coarseness**. It finds out the presence of any texture in an image and is measured by the size of the primitives which form the texture. Generally, a coarse texture comprises large sized primitives which are typified by the degree of neighboring uniformity of gray-levels. On the other hand, fine texture can be defined by small primitives and these are described by the degree of neighboring variations of gray-levels.

$$F_{cos} = \left( \epsilon + \sum_{i=0}^{L-1} p_i g(i) \right)^{-1} \tag{3}$$

where, $\varepsilon$ is a small number which avoids the coarseness coefficient to become infinite and $p_i$ is the estimated probability of the occurrence of the intensity values $i$ such that

$$p_i = N_i / n \tag{4}$$

with $N_i$ denoting the number of pixels having level $i$, and $n = (N - K)(M - K)$.

- **Contrast**. It quantifies the amount of clarity with which the different primitives in a texture can be differentiated. A well contrasted image is defined by the primitives which are visible as well as distinguishable. Among the factors that influence contrast, the gray-levels, the ratio of white and black pixels and the frequency of intensity changes of gray-levels are important.

$$F_{con} = \left[ \frac{1}{N_t(N_t - 1)} \sum_{i=0}^{L-1} \sum_{j=0}^{L-1} p_i p_j (i-j)^2 \right] \left[ \frac{1}{n} \sum_{i=0}^{L-1} g(i) \right] \qquad (5)$$

- **Busyness**. It measures the change of intensity from any pixel to its locality. If the intensity changes are quick and rush then it is called busy texture, whereas if the same are slow and gradual then it is called a non-busy texture. There is a relationship of busyness with the spatial frequency of the intensity changes in an image. Along with that, busyness is also affected by the amplitude of the intensity changes.

$$F_{bsuy} = \frac{\sum_{i=0}^{L-1} p_i g(i)}{\sum_{i=0}^{L-1} \sum_{j=0}^{L-1} |i p_i - j p_j|} \forall p_i \neq 0, \ p_j \neq 0 \qquad (6)$$

- **Complexity**. This is the visual information of texture. A texture is said to be complex when its information content is very high. This depends on the number of diverse primitives and average intensity values. Complexity is the sum of normalized differences between intensity values measured in pairs. These are weighted by the sum of the elements in the NGTDM corresponding to any two intensity values. Mathematically, it can be written as:

$$F_{com} = \sum_{i=0}^{L-1} \sum_{j=0}^{L-1} \frac{|i-j|}{n(p_i + p_j)} \left[ p_i g(i) + p_j g(j) \right] \forall p_i \neq 0, \ p_j \neq 0 \qquad (7)$$

- **Texture Strength**. Strength integrates and summarizes the concepts of busyness and coarseness. An image with a strong texture is composed by easily definable and clearly visible elements. It can be expressed as:

$$F_{str} = \frac{\sum_{i=0}^{L-1} \sum_{j=0}^{L-1} (p_i + p_j)(i-j)^2}{\epsilon + \sum_{i=0}^{L-1} g(i)} \forall p_i \neq 0, \ p_j \neq 0 \qquad (8)$$

For feature extraction purpose, each of the text-line images, written in different scripts, are firstly divided into 4 sub-images using 2-level quad tree decomposition approach and the five features are then computed from each of these sub-images. Two distances $d = 1$ and $d = 2$ are used in feature computation, corresponding to neighborhood sizes of $3 \times 3$ and $5 \times 5$ respectively. So, a feature vector of size 40 (F1-F40) is extracted for each text-line images using NGTDM. In the computation of $F_{cos}$ and $F_{str}$, the value of $\epsilon$ is taken as $10^{-7}$.

## 4.2  Gray-Level Run Length Matrix (GLRLM)

The application of a run length matrix for the purpose of texture feature extraction is proposed by Galloway [26]. Let, there is a given image of size $M \times N$, then a run-length matrix $p(i, j)$ is determined as the number of runs of pixels having gray-level $i$ and run length $j$.

- Short Run Emphasis (SRE):

$$SRE = \frac{1}{n_r} \sum_{i=1}^{M} \sum_{j=1}^{N} \frac{p(i,j)}{j^2} = \frac{1}{n_r} \sum_{j=1}^{N} \frac{p_r(j)}{j^2} \tag{9}$$

- Long Run Emphasis (LRE):

$$LRE = \frac{1}{n_r} \sum_{i=1}^{M} \sum_{j=1}^{N} p(i,j) \cdot j^2 = \frac{1}{n_r} \sum_{j=1}^{N} p_r(j) \cdot j^2 \tag{10}$$

- Gray-Level Non-uniformity (GLN):

$$GLN = \frac{1}{n_r} \sum_{i=1}^{M} \left( \sum_{j=1}^{N} p(i,j) \right)^2 = \frac{1}{n_r} \sum_{i=1}^{M} [p_g(i)]^2 \tag{11}$$

- Run Length Non-uniformity (RLN):

$$RLN = \frac{1}{n_r} \sum_{j=1}^{N} \left( \sum_{i=1}^{M} p(i,j) \right)^2 = \frac{1}{n_r} \sum_{j=1}^{N} [p_r(j)]^2 \tag{12}$$

- Run Percentage (RP):

$$RP = \frac{n_r}{n_p} \tag{13}$$

In the above equations, $n_r$ is the number of runs whereas $n_p$ is the number of pixels in the image. It is noticed that most of the features are only functions of $p_r(j)$, which do not consider the gray-level information of $p_g(i)$. Chu et al. [27] estimated two features to calculate gray-level information in the matrix.

- Low Gray-Level Run Emphasis (LGRE):

$$LGRE = \frac{1}{n_r} \sum_{i=1}^{M} \sum_{j=1}^{N} \frac{p(i,j)}{i^2} = \frac{1}{n_r} \sum_{i=1}^{M} \frac{p_g(i)}{i^2} \tag{14}$$

- High Gray-Level Run Emphasis (HGRE):

$$HGRE = \frac{1}{n_r} \sum_{i=1}^{M} \sum_{j=1}^{N} p(i,j) \cdot i^2 = \frac{1}{n_r} \sum_{i=1}^{M} p_g(i) \cdot i^2 \qquad (15)$$

Further, Dasarathy et al. [28] described another four feature estimation functions based on the concept of combined statistical measure of gray-level and run length, as follows:

- Short Run Low Gray-Level Emphasis (SRLGE):

$$SRLGE = \frac{1}{n_r} \sum_{i=1}^{M} \sum_{j=1}^{N} \frac{p(i,j)}{i^2 \cdot j^2} \qquad (16)$$

- Short Run High Gray-Level Emphasis (SRHGE):

$$SRHGE = \frac{1}{n_r} \sum_{i=1}^{M} \sum_{j=1}^{N} \frac{p(i,j) \cdot i^2}{j^2} \qquad (17)$$

- Long Run Low Gray-Level Emphasis (LRLGE):

$$LRLGE = \frac{1}{n_r} \sum_{i=1}^{M} \sum_{j=1}^{N} \frac{p(i,j) \cdot j^2}{i^2} \qquad (18)$$

- Long Run High Gray-Level Emphasis (LRHGE):

$$LRHGE = \frac{1}{n_r} \sum_{i=1}^{M} \sum_{j=1}^{N} p(i,j) \cdot i^2 \cdot j^2 \qquad (19)$$

These features are all based on intuitive reasoning, in an attempt to capture some apparent properties of run-length distribution. For each of the 11 measurements, defined above, the values of $\theta \in 0°, 45°, 90° \text{ and } 135°$ lead to a total of 44 (F41–F84) features using GLRLM. Finally, a set of 84 (i.e. 40 + 44) statistical textural features are extracted using both NGTDM and GLRLM for the text-line level classification of twelve different handwritten scripts.

## 5 Experimental Evaluation and Discussion

The performance of the present script identification scheme is evaluated on a dataset of 2400 preprocessed text-line images as described in Sect. 3. For each 200 text line images of a particular script, 135 images are applied for training and the rest 65 images are applied for testing purpose. Seven well-known classifiers *namely*, Naïve

**Table 1** Recognition performances of the proposed script identification technique using seven well-known classifiers (best case is shaded in gray and styled in bold)

| | Classifiers | | | | | | |
|---|---|---|---|---|---|---|---|
| | Naïve Bayes | Bayes Net | MLP | SVM | Random Forest | Bagging | MultiClass Classifier |
| Success Rate (%) | 89.74 | 90.95 | **97.69** | 95.87 | 94.6 | 91.18 | 93.37 |
| 95% confidence score (%) | 91.19 | 93.67 | **99.85** | 97.7 | 97.39 | 93.83 | 95.52 |

Bayes, Bayes Net, MLP, Support Vector Machine (SVM), Random Forest, Bagging and MultiClass Classifier are used to select the best classifier suitable for the present experimental setup. The recognition performances and their corresponding scores achieved at 95% confidence level are shown in Table 1.

As observed from Table 1 that MLP classifier produces the highest identification accuracy of 97.69%. In the present work, detailed error analysis of MLP classifier with respect to some well-known statistical parameters *namely*, Kappa statistics, Mean Absolute Error (MAE), Root Mean Square Error (RMSE), True Positive Rate (TPR), False Positive Rate (FPR), Precision, Recall, F-measure, Matthews Correlation Coefficient (MCC) and Area Under ROC (AUC) are also computed. The values of Kappa statistics, mean absolute error, root mean square error of MLP classifier for the present technique are found to be 0.9748, 0.0056 and 0.0557 respectively. Table 2 provides a statistical performance analysis of the remaining parameters for each of the aforementioned scripts.

**Table 2** Statistical performance measures along with their respective means (shaded in gray and styled in bold) achieved by the proposed technique for twelve handwritten scripts

| Scripts | TPR | FPR | Precision | Recall | F-Measure | MCC | AUC |
|---|---|---|---|---|---|---|---|
| *Bangla* | 0.812 | 0.000 | 1.000 | 0.812 | 0.896 | 0.893 | 0.906 |
| *Devanagari* | 0.990 | 0.001 | 0.990 | 0.990 | 0.990 | 0.989 | 0.999 |
| *Gujarati* | 0.990 | 0.004 | 0.962 | 0.990 | 0.976 | 0.973 | 1.000 |
| *Gurumukhi* | 1.000 | 0.005 | 0.953 | 1.000 | 0.976 | 0.974 | 0.999 |
| *Kannada* | 1.000 | 0.003 | 0.971 | 1.000 | 0.985 | 0.984 | 1.000 |
| *Malayalam* | 0.990 | 0.001 | 0.990 | 0.990 | 0.990 | 0.989 | 0.999 |
| *Manipuri* | 1.000 | 0.000 | 1.000 | 1.000 | 1.000 | 1.000 | 1.000 |
| *Oriya* | 1.000 | 0.006 | 0.935 | 1.000 | 0.967 | 0.964 | 1.000 |
| *Tamil* | 1.000 | 0.005 | 0.953 | 1.000 | 0.976 | 0.974 | 1.000 |
| *Telugu* | 1.000 | 0.002 | 0.981 | 1.000 | 0.990 | 0.989 | 1.000 |
| *Urdu* | 1.000 | 0.000 | 1.000 | 1.000 | 1.000 | 1.000 | 1.000 |
| *Roman* | 0.940 | 0.000 | 1.000 | 0.940 | 0.969 | 0.967 | 0.980 |
| **Weighted Average** | **0.977** | **0.002** | **0.978** | **0.977** | **0.976** | **0.975** | **0.990** |

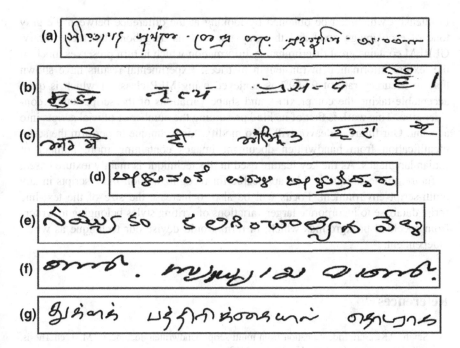

**Fig. 2** Sample text-line images written in **a** *Bangla*, **b** *Devanagari*, **c** *Gurumukhi*, **d** *Kannada*, **e** *Telugu*, **f** *Malayalam* and **g** *Tamil* misclassified as *Gujarati*, *Gurumukhi*, *Devanagari*, *Telugu*, *Kannada*, *Tamil* and *Malayalam* scripts respectively

Though Table 2 shows impressive results but some misclassifications have been found during the experimentation. The main reasons are: (a) presence of speckled noise, (b) existence of multi-skewed words present in some text-lines, and (c) occurrence of irregular spaces within text words, punctuation symbols, etc. in the text–line images. The structural resemblance in the character set of some of the Matra based scripts like *Devanagari* and *Gurumukhi* and non-Matra based scripts like *Kannada* and *Telugu* as well as *Malayalam* and *Tamil* cause similarity in the contiguous pixel distribution which in turns misclassifies them among each other. Figure 2 shows some samples of misclassified text-line images.

## 6  Conclusion and Future Work

We have proposed a robust method for handwritten script identification at text-line level for all the official scripts of India. The main intention of this paper is to facilitate the multilingual handwritten OCR and script based retrieval of offline handwritten documents. A set of 84 features are extracted using the combination of NGTDM and GLRLM. NGTDM aims to extract information about spatial changes in intensity which can be obtained by looking at the difference between the gray tone of each image pixel and the gray tones of its neighbors. On the contrary, GLRLM contains great discriminatory information which in turn preserves much of the texture information in run-length matrices. Experimental results have shown that an accuracy rate of 97.69% is achieved using MLP classifier which is quite acceptable taking the complexities and shape variations of the scripts under consideration. This work is first of its kind presuming the number of official scripts into account. Our future endeavor will be to modify this technique to perform the script identification from handwritten document images containing more number of Indian languages. As the key feature used in this technique is mainly texture based, in future, the technique will be applicable for recognizing non-*Indic* scripts in any multi-script environment. Focus will be also to increase the size of the text-line script database to incorporate larger variations of writing styles belonging to writers from speckled backgrounds which, in turn, would devise our technique as writer independent.

# References

1. Singh, P.K.: Script identification from multi-script handwritten documents. M. Tech Thesis, CSE Department, Jadavpur University (2013)
2. Language in India. http://www.languageinindia.com/feb2011/vanishreemastersfinal.pdf. Accessed 05 Feb 2016
3. Singh, P.K., Sarkar, R., Nasipuri, M.: Offline script identification from multilingual indic-script documents: a state-of-the-art. Comput. Sci. Rev. **15–16**, 1–28 (2015)
4. Dhandra, B.V, Nagabhushan, P., Hangarge, M., Hegadi, R.: Script identification based on morphological reconstruction in document images. In: IEEE International Conference of Pattern Recognition, Hong Kong, pp. 950–953 (2006)
5. Padma, M.C., Vijaya, P.A.: Global approach for script identification using wavelet packet based features. Int. J. Signal Process. Image Process. Pattern Recogn. **3**, 29–40 (2010)
6. Padma, M.C., Vijaya, P.A.: Wavelet packet based texture features for automatic script identification. Int. J. Image Process. **4**, 53–65 (2010)
7. Pal, U., Chaudhuri, B.B.: Identification of different script lines from multi-script documents. Image Vis. Comput. **20**, 945–954 (2002)
8. Padma, M.C., Vijaya, P.A.: Identification of Telugu, Devnagari and English scripts using discriminating features. Int. J. Comput. Sci. Inf. Technol. **1** (2009)
9. Padma, M.C., Vijaya, P.A.: Script identification from trilingual documents using profile based features. Int. J. Comput. Sci. Appl. **7**, 16–33 (2010)
10. Joshi, G.D., Garg, S., Sivaswamy, J.: Script identification from Indian documents. Lect. Notes Comput. Sci. (including Subser. Lect. Notes Artif. Intell. Lect. Notes Bioinform.) LNCS **3872**, 255–267 (2006)
11. Jindal, M., Hemrajani, N.: Script identification for printed document images at text-line level using DCT and PCA. IOSR J. Comput. Eng. **12**, 97–102 (2013)
12. Pal, U., Chaudhuri, B.B.: Automatic separation of words in multi lingual multi script indian documents. In: 4th International Conference on Document Analysis and Recognition (ICDAR). pp. 576–579 (1997)
13. Sinha, S., Pal, U., Chaudhuri, B.B.: Word-wise script identification from Indian documents. LNCS **3163**, 310–321 (2004)

14. Hassan, E., Garg, R., Chaudhury, S., Gopal, M.: Script based text identification : a multi-level architecture. In: Proceedings of the 2011 Joint Workshop on Multilingual OCR and Analytics for Noisy Unstructured Text Data, pp. 11:1–11:8 (2011)
15. Dhandra, B.V, Mallikarjun, H., Hegadi, R., Malemath, V.S.: Word-wise script identification from bilingual documents based on morphological reconstruction. In: IEEE International Conference on Digital Information Management, pp. 389–394 (2006)
16. Pati, P.B., Ramakrishnan, A.G.: Word level multi-script identification. Pattern Recogn. Lett. **29**, 1218–1229 (2008)
17. Dhanya, D., Ramakrishnan, A.G., Pati, P.B.: Script identification in printed bilingual documents. Sadhana Acad. Proc. Eng. Sci. **27**, 73–82 (2002)
18. Singh, P.K., Dalal, S.K., Sarkar, R., Nasipur, M.: Page-level script identification from multi-script handwritten documents. In: 3rd IEEE International Conference on Computer, Communication, Control and Information Technology (C3IT), pp. 1–6 (2015)
19. Hangarge, M., Dhandra, B.V: Offline handwritten script identification in document images. Int. J. Comput. Appl. **4** (2010)
20. Singh, P.K., Sarkar, R., Nasipuri, M.: Line-level script identification for six handwritten scripts using texture based features. In: 2nd Information Systems Design and Intelligent Applications, Advances in Intelligent Systems and Computing, pp. 285–293 (2015)
21. Roy, K., Pal, U.: Word-wise Handwritten Script Separation for Indian postal automation. In: International Workshop on Frontiers in Handwriting Recognition, La Baule, pp. 521–526 (2006)
22. Sarkar, R., Das, N., Basu, S., Kundu, M., Nasipuri, M., Basu, D.K.: Word level script identification from Bangla and Devnagari handwritten texts mixed with Roman scripts. J. Comput. **2**, 103–108 (2010)
23. Singh, P.K., Sarkar, R., Das, N., Basu, S., Nasipuri, M.: Identification of Devnagari and Roman scripts from multi-script Handwritten documents. In: 5th International Conference on Pattern Recognition and Machine Intelligence (PReMI), pp. 509–514 (2013)
24. Gonzalez, R.C., Woods, R.E.: Digital Image Processing. Prentice-Hall, India (1992)
25. Amadasun, M., King, R.: Textural features corresponding to textural properties. IEEE Trans. Syst. Man Cybern. **19**, 1264–1274 (1989)
26. Galloway, M.M.: Texture analysis using gray level run lengths. Comput. Graph. Image Process. **4**, 172–179 (1975)
27. Chu, A., Sehgal, C.M., Greenleaf, J.F.: Use of gray value distribution of run lengths for texture analysis. Pattern Recogn. Lett. **11**, 415–420 (1990)
28. Dasarathy, B.R., Holder, E.B.: Image characterizations based on joint gray-level run-length distributions. Pattern Recogn. Lett. **12**, 497–502 (1991)

# An Approach to Stroke-Based Online Handwritten Bangla Character Recognition

Shibaprasad Sen, Ram Sarkar and Kaushik Roy

**Abstract** This paper deals with stroke-based online Bangla character recognition strategy. In the present work, constituent strokes have been extracted from characters and then popularly used distance based features have been estimated in order to recognize the basic strokes. Next, a rule-based approach is followed for the recognition of the characters from the previously recognized strokes. A total of 15,000 isolated online handwritten Bangla characters contributing 32,534 stroke samples have been used in this experiment, and a satisfactory result of 89.39% recognition accuracy has been achieved.

**Keywords** Online character recognition · Stroke based classification · Distance based feature · Rule-based approach · Bangla script

## 1 Introduction

Exponentially increasing usage of handheld devices such as Take Note, iPad, Smart phones etc. by individuals makes Online Handwriting Recognition (OHR) an upcoming research domain. The advantage of using such devices is that people from major part of the society can write information freely on those devices in their natural handwriting style. These written data are saved in the form of online information. Writing in normal style on those devices not only saves extra time but also minimizes the chance of mistyping that may arise when writing with a

S. Sen (✉)
Future Institute of Engineering & Management, Kolkata, India
e-mail: shibubiet@gmail.com

R. Sarkar
Jadavpur University, Kolkata, India
e-mail: raamsarkar@gmail.com

K. Roy
West Bengal State University, Barasat, India
e-mail: kaushik.mrg@gmail.com

© Springer Nature Singapore Pte Ltd. 2017     153
R. Chaki et al. (eds.), *Advanced Computing and Systems for Security*,
Advances in Intelligent Systems and Computing 568,
DOI 10.1007/978-981-10-3391-9_10

keyboard. Thereby, researchers are now showing their interest to OHR. Though good number of research works is present in the literature for Devanagari script [1–8] but for Bangla script, research is not still mature enough. Bhattacharya et al. in [9] proposed a novel direction code based online Bangla character recognition with a database of size 7043. In this approach authors divided the online temporal data sequence of specimen character into several sub-divisions and then computed the direction code feature into these sub-divisions. They made a histogram of these produced values and used them as feature for their experiment. Authors in [10] manually grouped the strokes for all specimens of 50 Bangla character samples into 54 classes. Grouping into different classes was done by considering the shape similarity of the graphemes constituting character sample. They used Hidden Markov Model (HMM) for stroke recognition. One HMM was used for each stroke class. An unknown character was then identified by looking into look up tables for the produced strokes. A stroke based approach was adapted in [11] for the recognition of online handwritten Bangla characters where strokes were represented as a string of shape features. Dynamic Time Wrapping (DTW) algorithm was applied to recognize the component strokes from a stroke database. After identifying all the constituent strokes finally the characters were recognized. A different way was discussed in [12] by Roy where strokes were collected from online character information and then stroke level sequential and dynamic information were collected on the basis of pen movements on those writing devices. This sequential and dynamic information were serving as feature values for the recognition of strokes in his work. Bhattacharya et al. in [13] engaged themselves for the annotation of unconstrained online handwritten Bangla samples and thereby prepared a substantial volume of annotated dataset for those handwriting specimens. A graphical user interface based semi-automatic system was built for the annotation at the character boundary level. Authors in [14] proposed a two-stage approach towards online Bangla character recognition where firstly probability distribution was estimated for each stroke class by using some stroke level features. A HMM based character classifier was then constructed by considering each stroke class as a state. Authors in [15], proposed a new scheme where the constituent strokes were divided into a number of local zones. Stroke level structural and directional information were computed in each local zone and then these computed values were concatenated to serve as feature values for the experiment. Mondal et al. in [16] applied direction code histogram and point-float feature extraction techniques towards recognition of Bangla alphabets. Effectiveness of the feature values were tested by HMM, Nearest Neighbor and Multilayer Perceptron (MLP) classifiers. Authors in [17] combined some online feature (namely, point based and structural) and some offline (quad tree based longest run and convex hull) features for online Bangla characters recognition. Sen et al. in [18] explored a simple but effective feature computation strategy for the recognition of online Bangla basic characters known as distance based feature. Authors in [19] dealt with the individual impact of Global Information and Local Information based feature extraction approaches towards online Bangla character recognition. Global Information monitored the shape statistics, whereas, Local Information tried to extract some valuable local information about the Bangla character samples. Authors in this paper also highlighted the combining impact of Global and Local Information based techniques. If number of speakers would be a key criterion for more research, then Bangla language definitely demands more attention from the OHR research community. In the current work, a new stroke-based approach has been adapted to recognize the online handwritten Bangla characters. In this approach, distance based features have been computed for the recognition of constituent strokes of the specimen character samples. This feature vector is then feed to MLP classifier for the recognition of the strokes. At the end a rule-based approach is constructed for the recognition of target character from component strokes.

In this paper, the more concentration is given towards the feature extraction stage that generally is the key factor in pattern recognition. Researchers face difficulties to recognize the online handwritten Bangla characters due to the facts such as the presence of moderately large volume of character and strokes symbol set, extremely cursive nature of the shapes of characters even if written isolated, and there exist few groups of almost similar shape strokes or characters. Inherent problem for any online handwriting system occurs due to writing style of different individual in terms of shape and size of the symbols; even same writer writes differently at different times. Especially in Bangla, the number and order of strokes of an online handwritten character make the thing complex enough.

## 2 Bangla Script

Bangla is known as the second most popular language in India. It is also one of the official languages in India as well as an official script in Bangladesh. It is inherited from ancient Brahmi script though its exact derivation is disputed. Writing style has some common similarities with Dravidian scripts, especially in the shapes of some vowels, but it is more similar to the Indo-Aryan scripts, in particular Devanagari. In Bangla, every consonant represents a syllable containing an inherent vowel written from left to right. Basic Bangla character set contains 39 consonants and 11 vowels.

## 3 Working Methodology

Any stroke-based online handwriting recognition system follows some basic steps, such as (a) strokes are extracted from characters and then some pre-processing steps are required, where at first smoothing is done to remove noise and duplicate (or repeating points) points are discarded. Then pixel points forming the input strokes are re-sampled to obtain a new sequence of points, which are approximately equidistant. All the stroke data sample is then normalized into a fixed number of points and scaling is performed to fix the sample into a pre-defined window size, (b) after the pre-processing stage features are extracted from those pre-processed data, (c) the stroke samples are recognized through some well-known classifier(s) based on feature values extracted from the former stage and (d) finally, characters are identified from recognized strokes.

## 3.1 Stroke Database Generation

Data collection sheets have been arranged for the preparation of isolated character database as it is not feasible to collect the strokes at random. This is because normal human are not acquainted with writing strokes individually. Here the strokes are extracted from their parent characters. 300 handwritten specimens of each Bangla character sample have been collected from 100 different persons. Different sections of society with varying age groups, educational background, gender etc. have contributed in the data collection drive. As Bangla basic alphabet set contains 50 distinct character shapes, hence in the present experiment, 15,000 data samples in total have been collected. No burden has been imposed on the writers except that the writers were requested to write constituent strokes as part of basic stroke database (see Fig. 1). Take note was used to collect data which can be represented as a collection of pen position $p_t$, where $t$ is ranging from 1 to $M$. $p_t$ describes pen position having $x$ coordinate as $x_t$ and $y$ coordinate $y_t$ with pen up or pen down status. Here, $M$ represents total pen positions for each of the character samples. Stroke can be described by some sequence of pen positions collected between one pen-down and pen-up. Bangla characters are generally formed by combination of one or more basic strokes. From statistical analysis on the dataset it is found that the minimum number of strokes used to write a Bangla character is one and maximum number for the same is six. For example এ, ও and ৎ can be written using a single

| Bangla Stroke | Nomenclature | Bangla Stroke | Nomenclature | Bangla Stroke | Nomenclature | Bangla Stroke | Nomenclature |
|---|---|---|---|---|---|---|---|
| ৩ | 3_: | এ | E__ | ঞ | JA_ | ⌒ | PA2 |
| ↲ | _AA | ও | O__ | ঃ | EN_ | ⁊ | PA3 |
| — | MTR | ☞ | KA_ | ৬ | DDH | ২৸ | PHA |
| ‖ | _A_ | ১ | 1__ | ৳ | T__ | ৬ | BHA |
| ‖T | AAA | ২ৰ | KH_ | ৯ | TTA | ২↘ | MA_ |
| ২ | 2__ | শা | G__ | ৯ | TT_ | ২৷ | YA_ |
| ⁓ | I__ | ৮ | G1_ | ৫ | MN_ | ০ | 0__ |
| ᒲ | II_ | ৲ | G2_ | ৩ | MN1 | ৴ | LA_ |
| ৬ | 6__ | ৰ৷ | GH_ | ২ৰ | THA | ২ৰ | SA_ |
| ↙ | UU_ | ৩ | UN_ | ↳ | DA_ | ↳ | MSA |
| ৰ৷ | DHA | ৮ | C__ | ৎ | DH1 | ২৸ | S__ |
| ৰ | BA_ | ৩ | CH_ | ↲ | NA_ | ৎ | S1_ |
| ৸৸ | RI_ | ৲ | JA1 | ৴ | PA1 | ৎ | KT_ |

Fig. 1 Basic strokes of Bangla character set

stroke whereas to write উ an average of four strokes is needed. Maximum six strokes can be used for আ and rest of the characters bear an average stroke number that is less than three. As per observation, the average number of stroke per character is 2.2. The present database contains 50 basic symbols from Bangla character set and the total number of stroke for the Bangla script would be approximately 110. But from statistical analysis it is found that only 52 strokes are enough to represent all the basic characters. This is because some of the strokes are common to various characters.

From the Fig. 1 it can also be stated that 40 out of the 50 characters mostly contain a horizontal line at the upper part called Matra. The recognition of Bangla script is more difficult not only due to the presence of large volume of character set but also the variation in terms of number and order of strokes appear while writing the characters. To explain the situation, character 'আ' has been considered as an example because this character contains maximum number of strokes. Seven possible strokes coming from the specimens of this character sample are reflected in Fig. 2. It has also been observed that minimum two to maximum six strokes out of seven are used for writing the same character symbol. Some different ways with the varying number of strokes to write the character আ is shown in Fig. 3. Stroke order variation is depicted in Fig. 4, where four strokes have been considered to write আ.

**Fig. 2** Examples of strokes use to write আ

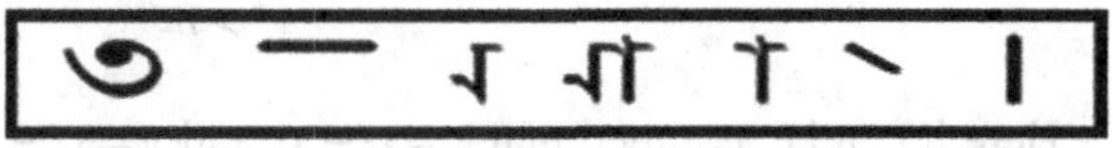

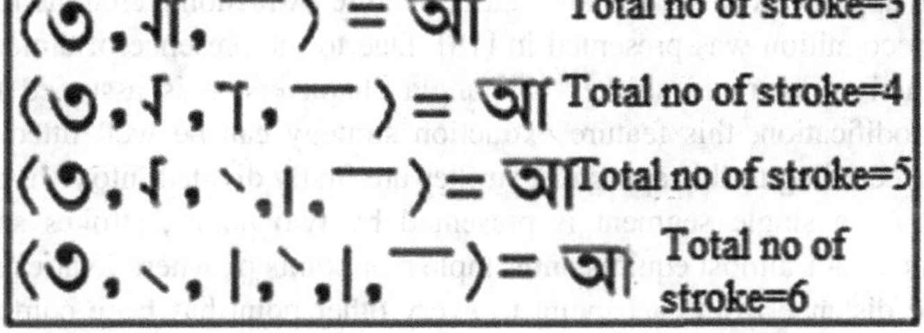

**Fig. 3** Some ways of writing আ using the possible stroke-combinations

**Fig. 4** Example of the
different stroke-order for a
character having four strokes

## 3.2  *Preprocessing and Feature Extraction*

### 3.2.1  Preprocessing

At pre-processing step, all the duplicate or repeated points have been removed at first from the stroke specimen because those points may increases redundancy with added computational complexity. If $p_i$ and $p_k$ refer to two consecutive pen points, then ith point $p_i$ is retained with respect to kth point $p_k$ if the Eq. (1) is maintained.

$$x^2 + y^2 > m^2 \tag{1}$$

Here it can be assumed that $x = x_i - x_k$ and $y = y_i - y_k$. The value of $m$ is set to 0 to remove all duplicate points. Next, pixel points constituting the strokes are re-sampled to generate a new sequence of positions that are approximately equidistant. Then the stroke is normalized into 64 points by keeping the stroke structure intact. When performing scaling operation all normalized points are fixed to a predefined window of size $512 \times 512$.

### 3.2.2  Distance Based Feature

A different approach toward Distance based feature extraction technique for Bangla character recognition was presented in [18]. Due to the presence of almost similar kind of cursive nature in strokes of Bangla characters, it is assumed that with suitable modification, this feature extraction strategy can be well fitted into this problem. According to this approach, strokes are firstly divided into $N$ hypothetical segments. As a single segment is presented by two points, strokes shapes are divided into $N + 1$ almost equidistant sample pen points $p_i$, where $i$ varies from 1 to $N + 1$. The distance from each point to every other point has been computed and these distance values are considered as the features for the current experiment (see Fig. 5). According to statistical analysis of Bangla stroke set, it has been noticed that there exists some strokes those are nearly similar in shape structure. To defend this situation, some local information of the stroke are required which in turn

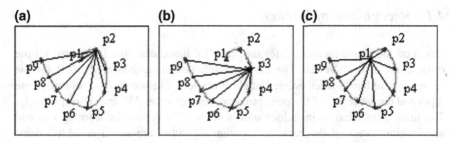

**Fig. 5 a–c** Distance calculation from one segmented point to the rest (*red points* refer to segmented points and *black lines* denote the distances among segmented points)

produce discriminating features for recognition of such strokes. In the current work, all strokes are divided into 16 segments. It is really challenging to decide the optimal number of segmentations needed but we have followed the work as mentioned in [18], and have set the value of N to 16. Distance based feature algorithm is presented in Algorithm 1.

```
Step 1: START
Step 2: i=1;
Step 3: j=i+1;
Step 4: compute the distance d between pᵢ and pⱼ;
Step 5: j=j+1;
Step 6: if (j<=N+1) then goto STEP 4,
otherwise goto STEP 7;
Step 7: i=i+1;
Step 8: if (i<N+1)goto STEP 3,otherwisegoto STEP 9;
Step 9: END.
```

**Algo 1.** Algorithm used to compute distance based features

Number of features generated by distance based feature extraction approach for N segmented stroke is $N*(N+1)/2$. As in the present work N = 16, so a 136-element feature vector has been generated for recognizing the basic strokes.

# 4 Recognition

In this work, the recognition module has been divided into two parts: (i) Recognition of strokes and (ii) Formation of valid character from recognized strokes.

## 4.1 Recognition of Strokes

Based on the above mentioned feature, an MLP based classifier is adopted for the recognition of the strokes. The current work designs a 2-layered MLP for the recognition of handwritten online Bangla characters. The numbers of neurons in the input and output layers of the perceptron are considered as 136 and 52 respectively. The number of neurons in hidden unit is set to 85 whereas the learning rate and acceleration factor of the back propagation algorithm are set to suitable values, based on trial runs. A network of 136-85-52 is thereby designed.

## 4.2 Formation of Valid Character from Recognized Strokes

Each character will be constructed with the help of its recognized strokes. To do so, firstly, after extracting all the strokes from all the characters, strokes are classified into three categories depending on positions of the same.

Major Stroke: A stroke is called Major stroke if it occupies major portion of the character symbol. For the character 'অ' the major stroke is 'ও'.
Minor Stroke: A stroke is called Minor stroke if it occupies minor portion of the character. For the character 'অ' the minor stroke is 'া'.
Matra(MTR): A stroke is called Matra if it is a horizontal straight line occurred at the upper part of the character. For the character 'অ' the matra stroke is '-'.

This categorization plays an important role at the time of forming characters from the identified strokes. To minimize the problem that may arise due to different sequence of stroke-order even in case of same character, Major stroke and Matra are identified and they are placed at first and last positions respectively in stroke sequence. This reduces the possible number of different stroke-order combinations while forming the character. A rule-based approach is then constructed to form all such stroke-order combinations. Few such rules for recognizing characters অ and আ are shown in Fig. 6.

| stroke1 ▾ | stroke2 ▾ | stroke3 ▾ | stroke4 ▾ | Result ▾ |
|---|---|---|---|---|
| 3__ | _AA | - | - | A |
| 3__ | _AA | MTR | - | A |
| 3__ | AAA | - | - | AA |
| 3__ | AAA | MTR | - | AA |
| 3__ | _AA | _A_ | - | AA |
| 3__ | _AA | _A_ | MTR | AA |

**Fig. 6** Rules for constructing characters অ and আ

# 5  Results and Discussion

The experimental evaluation of the above technique has been carried out using isolated Bangla strokes. A total of 32,534 strokes are prepared from the collected online handwritten Bangla basic characters, as mentioned earlier.

## 5.1  Recognition Result on Isolated Stroke and Formation of Characters

In the present work, MLP classifier produces 89.39% accuracy in recognizing strokes of online Bangla basic characters. Distance based feature exhibits reasonable performance to distinguish different cursive structural patterns for recognizing basic strokes except some shown in Table 1. From this table it can be seen that the stroke MTR sometimes misclassified as MSA or as _A_ in the first row. MTR, MSA, _A_ are structurally similar but they differ only by their directions. For example, the MTR is almost horizontal in nature whereas MSA has a small slope along x axis, and, _A_ is vertical in nature. Similarly, strokes UU_ and I__ (see the second row of Table 1) are misclassified for the same reason. First few portions (cursive part) for both the strokes in third row are same but the next portion is either vertical or horizontal. Strokes in the fourth row are structurally almost similar.

Authors in [10] used shape based feature for stroke recognition and achieved 84.6% accuracy. In [14], some structural and directional features were used for stroke recognition with 87.48% success rate, whereas 5000 test samples used. Roy et al. in [12] have used point based with some structural features toward stroke classification. Their technique classified 96.85% strokes accurately. In contrast, the present technique yields better performance than [10, 14]. Though this approach exhibits lesser accuracy than [12] but it solves certain problems faced by them. Hence, it can be concluded that distance based feature used here is quite capable of

**Table 1**  Most confusing stroke pairs

| Original stroke | Misclassified as | |
|---|---|---|
| MTR( ➖ ) | (MSA) ＼ | ( A_ ) ˧ |
| (I_ ) ↩ | (UU_ ) ⌣ | |
| (NA_ ) ⩗ | (2_ ) ⩨ | |
| (C_ ) ♭ | (DDH) ♭ | |

**Table 2** Most confusing
character pairs

| Original character | Misclassified as |
|---|---|
| চ | ড |
| য | ষ |

identifying the cursive structure accurately. Moreover, the datasets used in the works [10, 12, 14] are different thereby these works cannot be compared directly with the present work. But to facilitate the researchers in this domain, the facts are highlighted in this section.

Recognition of handwritten online Bangla basic characters from constituent strokes using rule-based technique heavily depends on stroke recognition results. Table 2 reflects some misclassified character pairs. After a detailed analysis, it is found that characters are misclassified because of error in stroke classification stage and strong structural similarity of the stroke samples. In first row, character য is misclassified as ষ and vice versa. The main reason for this is that in both the samples, major stroke is same but when MTR is misclassified as MSA then character য is recognized as ষ and when MSA is misclassified as MTR then character ষ is identified as য wrongly. These observations may be useful to explore more in the feature extraction stage for improving the overall recognition ability.

# 6   Conclusion

In the present work, distance based feature extraction approach is employed for recognition of strokes of the online handwritten Bangla characters. The outcome of this stage is applied to form the characters using a rule-based approach. Here, the success rate of stroke recognition plays an important to obtain a good character recognition system. There are still rooms to experiment with different number of segmentations with some novel features in order to identify the misclassified strokes correctly. Therefore, a need for corrective measure for misclassified strokes can be thought as the future scope of the present work.

# References

1. Connell, S.D., Sinha, R.M.K., Jain, A.K.: Recognition of unconstrained online Devanagari characters. In: 15th International Conference on Pattern Recognition, pp. 368–371 (2000)
2. Joshi, N., Sita, G., Ramakrishnan, A.G., Deepu, V.: Machine recognition of online handwritten Devanagari characters. In: Proceedings of International Conference on Document Analysis and Recognition, pp. 1156–1160 (2005)
3. Swethalakshmi, H., Jayaraman, A., Chakravarthy, V.S., Sekhar, C.C.: On-line handwritten character recognition for Devanagari and Telugu scripts using support vector machines. In: Proceedings of International Workshop on Frontiers in Handwriting Recognition, pp. 367–372 (2006)

4. Swethalakshmi, H., Sekhar, C.C., Chakravarthy, V.S.: Spatiostructural features for recognition of online handwritten characters in Devanagari and Tamil scripts. In: Proceedings of
   International Conference on Artificial Neural Networks, vol. 2, pp. 230–239 (2007)
5. Kumar, A., Bhattacharya, S.: Online Devanagari isolated character recognition for the iphone
   using hidden markov models. In: International Conference on Students Technology
   Symposium, pp. 300–304 (2010)
6. Tripathi, A., Paul, S.S., Pandey, V.K.: Standardization of stroke order for online isolated
   Devanagari character recognition for iPhone. In IEEE International Conference on
   Technology Enhanced Education, pp. 1–5 (2012)
7. Kubatur, S., Sid-Ahmed, M., Ahmadi, M.: A neural network approach to online Devanagari
   handwritten character recognition. In: International conference on High Performance
   Computing and Simulation (2012). doi:10.1109/HPCSim.2012.6266913
8. Lajish, V.L., Kopparapu, S.K.: Online handwritten Devanagari stroke recognition using
   extended directional features. In: IEEE 8th International Conference on Signal Processing and
   Communication System (2014). doi:10.1109/ICSPCS.2014.7021063
9. Bhattacharya, U., Gupta, B.K., Parui, S.K.: Direction code based features for recognition of
   online Handwritten characters of Bangla. In: International Conference on Document Analysis
   and Recognition, pp. 58–62 (2007)
10. Parui, S.K., Guin, K., Bhattacharya, U., Chaudhuri, B.B.: Online handwritten Bangla
    character recognition using HMM. In International Conference on Pattern Recognition,
    pp. 1–4 (2008)
11. Bandyopadhyay, A., Chakraborty, B.: Development of online handwriting recognition
    system: a case study with handwritten Bangla character. In: World Congress on Nature and
    Biologically Inspired Computing, pp. 514–519 (2009)
12. Roy, R.: Stroke-database design for online handwriting recognition in Bangla. In:
    International Journal of Modern Engineering Research, pp. 2534–2540 (2012)
13. Bhattacharya, U., Banerjee, R., Baral, S., Dey, R., Parui, S.K.: A semi automatic annotation
    scheme for Bangla online mixed cursive handwriting samples. In: International Conference on
    Frontiers in Handwriting Recognition, pp. 680–685 (2012)
14. Biswas, C., Bhattacharya, U., Parui, S.K.: HMM based online handwritten Bangla character
    recognition using Dirichlet distributions. In: International Conference on Frontiers in
    Handwriting Recognition, pp. 600–605 (2012)
15. Ghosh, R.: A novel feature extraction approach for online Bengali and Devanagari character
    recognition. In: International Conference on Signal Processing and Integrated Networks,
    pp. 483–488 (2015)
16. Mondal, T., Bhattacharya, U., Parui, S.K., Das, K., Mandalapu, D.: On-line handwriting
    recognition of indian scripts—the first benchmark. In: 12th International Conference on
    Frontiers in Handwriting Recognition, pp. 200–205 (2010)
17. Sen, S.P., Paul, S.S., Sarkar, R., Roy, K., Das, N.: Analysis of different classifiers for on-line
    Bangla character recognition by combining both online and offline information. In: 2nd
    International Doctoral Symposium on applied computation and security Systems (2015)
18. Sen, S.P., Sarkar, R., Roy, K.: A simple and effective technique for online handwritten Bangla
    character recognition. In: 4th International Conference on Frontiers in Intelligent Computing:
    Theory and Application, pp. 201–209 (2015)
19. Sen, S.P., Bhattacharyya, A., Das, A., Sarkar, R., Roy, K.: Design of novel feature vector for
    recognition of online handwritten Bangla basic characters. In: 1st International Conference on
    Intelligent Computing and Communication (2016)

# Functional Group Prediction of Un-annotated Protein by Exploiting Its Neighborhood Analysis in Saccharomyces Cerevisiae Protein Interaction Network

Sovan Saha, Piyali Chatterjee, Subhadip Basu and Mita Nasipuri

**Abstract** Identification of unknown protein function is important in biological field based on the fact that proteins are responsible for some vital diseases whose drug is still yet to be discovered. Protein interaction network serves a crucial role in protein function prediction among all the other existed methodologies. Motivated by this fact different neighborhood approaches are proposed by exploiting the various indispensable neighborhood properties of protein interaction network which has added an extra dimension to this field of study.

**Keywords** Protein interaction network · Protein function prediction · Functional groups · Neighbourhood analysis · Relative functional similarity · Edge clustering coefficient · Protein path connectivity · Node weight · Edge weight · Physico-chemical properties

S. Saha (✉)
Department of Computer Science and Engineering, Dr. Sudhir Chandra
Sur Degree Engineering College, Dumdum, Kolkata 700074, India
e-mail: sovan.saha@gmail.com; sovansaha12@gmail.com

P. Chatterjee
Department of Computer Science and Engineering, Netaji Subhash
Engineering College, Garia, Kolkata 700152, India
e-mail: chatterjee_piyali@yahoo.com

S. Basu · M. Nasipuri
Department of Computer Science and Engineering, Jadavpur University,
Kolkata 700032, India
e-mail: subhadip.basu@cse.jdvu.ac.in; subhadip@cse.jdvu.ac.in

M. Nasipuri
e-mail: mita.nasipuri@cse.jdvu.ac.in; mitanasipuri@yahoo.com

© Springer Nature Singapore Pte Ltd. 2017                                    165
R. Chaki et al. (eds.), *Advanced Computing and Systems for Security*,
Advances in Intelligent Systems and Computing 568,
DOI 10.1007/978-981-10-3391-9_11

# 1  Introduction

Determining unknown protein function prediction by biological experiments will lead to the expense of much more cost as well as time. Thus computational methodologies like gene fusions, gene neighborhood, protein-protein interactions (PPI) etc. are taken into consideration which not only reduce human effort, time and cost but also succeeded in generating much more accurate results evaluated through various standard metrics. In our work, we have considered PPI network for function prediction. An unknown protein may be annotated with a function of any protein in it's neighborhood as it is believed that neighborhood proteins perform almost similar function as that of the unknown one. Now the selection of neighborhood protein should be logically and computationally justified which is the main objective of all our proposed methods. Before proceeding into main section of our work, various existing methodologies have been discussed in the upcoming section in order to fetch a clear idea about the possible ways of unknown protein function prediction.

A simple neighborhood-counting method is proposed in the work of Schwikowski [1] for the assignment of functions of neighborhood to an unannotated protein depending on the most frequent occurrence of functional labels among its interacting partners. But in the chi-square method [2] function assignment depends on the largest *chi-square* scores which is usually defined as $\frac{(n_f - e_f)^2}{e_f}$, where $n_f$ denotes the number of proteins in the neighborhood of the unknown protein that have the function $f$ and $e_f$ denotes the expectation of this number based on the frequency of $f$ among all proteins in the network. Chen et al. [3] takes this neighborhood property to the next level by the inclusion of the concept of functional similarity between the neighbors from the level-1 and level-2 and the unknown protein in question. Vazquez et al. [4] generates an optimization problem using simulated annealing where they mainly highlight on maximizing unknown protein connectivity by increasing the number of interconnected edges with it. Karaoz et al. [5] applies an identical approach but to a large area of application including PPI data and gene expression data taking gene ontology in consideration. Some other approaches like flow based approach [6], markov random field [7], binomial model based loopy belief propagation [8], probabilistic methods [9], network based statistical algorithm [10] and UVCLUSTER based on bi-clustering [11] also emerge as successful and remarkable methodologies in PPI based protein function prediction. While on the other hand, Molecular Complex Detection (MCODE) [12] and graph clustering approaches [13, 14] also leave their marks in this field. Pruzli et al. [15] use graph theoretic approach to form clusters which are later analyzed by Highly Connected Sub-graphs (HCS) algorithm. King et al. [16] proposed application of Restricted Neighborhood Search Clustering algorithm (RNCS) for clustering the protein interaction network. Wei et al. [17] proposed a unique method of enriching the original PPI network by the incorporation of two type of edges: implicit and explicit thereby incorporating more information than the earlier one which will boost up the function prediction. Zhao et al. [18] creates a dynamic weighted interactome network from which prediction is executed using a ranking methodology.

The above study has revealed the fact that though existed methodologies have already created a larger impact but still there are lots of improvements yet to be incorporated like protein domain, structure and sequence related information. Motivated by this fact, various neighborhood based methods for the prediction of functions of unannotated proteins has been proposed in this work where the uncharacterized protein is associated with the function corresponding to the highest scoring neighborhood protein among all of them.

## 2 Dataset

We have used Munich Information Center for Protein Sequences database in our works [19–21]. The Munich Information Center for Protein Sequences (MIPS) is located at the Institute for Bioinformatics (IBI), which is part of the GSF-National Research Center for Environment and Health. The MIPS focuses on genome oriented bioinformatics, in particular the systematic analysis of genome information including the development and application of bioinformatics methods in genome annotation, expression analysis and proteomics. The database is incorporated with protein-protein interaction data of yeast (*Saccharomyces Cerevisiae*) which contains 15613 genetic and physical interactions. After discarding self-interactions a set of 12487 unique binary interactions involving 4648 proteins are selected.

## 3 Related Terminologies

A scoring scheme known as Neighborhood ratio ($P_{O_{l(=1....15)}}^{l(=1,2)}$) that computes the ratio of level-1 (or level-2) neighbors corresponding to a functional group and the total number of level-1 (or level-2) neighbors has been used in FunPred-0.1 and FunPred-0.2 [19]. Secondly, another amino acid scoring scheme involving protein sequence properties known as physico-chemical properties score ($PCP_{score}^{l(=1,2)}$) has been used in our next work [20]. In addition to these, other graph theory based measures like Protein Neighborhood Ratio Score ($Pscore^{l(=1,2)}$) [21], Relative functional similarity ($W_{u,v}^{l(=1,2)}$) [21, 22], Proteins path-connectivity score ($Q_{u,v}^{l(=1,2)}$) [21, 23] are used in FunPred-1 [21]. The higher the Protein Functional Similarity is, more functional similarity is found in those proteins. Protein path connectivity score measures a protein's connectivity in a network. Proteins of more paths and shorter path lengths are tightly connected in the network. Edge Clustering Coefficient is another graph theoretic measure which describes how close two proteins are. The protein having higher value of Edge Clustering Coefficient is likely to be more involved in the community. While in FunPred 2, we have used two important scoring schemes Edge weight ($W_{uv}$) [23] and Node weight ($W_v$) [23]. All the other relevant graphical terms and properties are described in our earlier works [19, 21].

## 4 Proposed Method

In this paper, three types of protein function prediction schemes, namely, FunPred-0, FunPred-1 and FunPred 2 are outlined where different aspects of neighborhood analysis are done. FunPred 0.1 and FunPred 0.2 [19] are very simpler methods which use neighborhood ratio of protein of unknown function. FunPred-0.1 selects proteins randomly from the protein interaction network and predicts the functions using level-1 and level-2 neighborhood ratios. The maximum neighborhood score among all its neighbors dominates the role to take decision for its functional group. FunPred-0.1 does this prediction trivially by considering level-1 and level-2 neighbors of entire sub graph. As a consequence, computation is done for every level- and level-2 neighbors including distantly related neighbors. This fact necessitates finding out the densely connected region instead of considering the entire neighborhood graph which helps to reduce computational time. In FunPred-0.2, identification of densely connected region is found from which relevant interactions of candidate protein is obtained. To do this an intelligent technique empowered with heuristic knowledge is applied in the protein interaction network. Then similar strategies followed in FunPred-0.1 are taken to predict the possible functional group of that candidate protein. The overall *match rate* achieved in method-I is 95.8% and in FunPred-0.2, it is 97.8% over 15 functional groups (cell cycle control ($O_1$), cell polarity ($O_2$), cell wall organization and biogenesis ($O_3$), chromatin chromosome structure ($O_4$), co-immuno-precipitation ($O_5$), co-purification ($O_6$), DNA Repair ($O_7$), lipid metabolism ($O_8$), nuclear-cytoplasmic transport ($O_9$), pol II transcription ($O_{10}$), protein folding ($O_{11}$), protein modification ($O_{12}$), protein synthesis($O_{13}$), small molecule transport ($O_{14}$) and vesicular transport ($O_{15}$)).

FunPred-1.1 and FunPred-1.2 [21] both are designed on the basis of FunPred-0 but they also take other graph theory based measures into account to predict function of an unknown protein. Here, less but more important functional groups are considered in this method. Eight functional groups are chosen as most of the proteins are found to be associated with these functional groups. Chosen functional groups are Cell cycle control, Cell Polarity, Cell wall organization and biogenesis, Chromatin chromosome structure, nuclear-cytoplasmic transport, poll-II transcription, Protein folding and Protein modification. For each functional group, 10% of proteins are chosen as candidate protein whose functional group has to be predicted whereas remaining 90% proteins are considered as their neighbors. Like FunPred-0, 2 types of neighbors like: level-1 neighbor (having direct connection with candidate protein) and level-2 neighbor (having direct connection with level-1 neighbors) are considered. Unlike FunPred-0, FunPred-1 computes neighborhood score than only computing neighborhood ratio. Neighborhood score is additive measure of neighborhood ratio, relative functional similarity and protein path connectivity score. For a candidate protein, neighborhood score is computed in level-1 and level-2 for each functional group. So FunPred-1.1 assigns the candidate protein to that functional group in which maximum neighborhood score is found. The overall accuracy of FunPred-1.1 is 75.8% and average recall and precision are evaluated as $0.5038 \pm 0.1810$ and $0.7613 \pm 0.1792$ respectively. The F-score is $0.6019 \pm 0.1770$. Like FunPred 0.2, FunPred-1.2 finds the densely connected region but using edge clustering coefficient of a protein. Edge clustering coefficient describes another neighborhood property stating how close two proteins are. However, if the computation is done only on significant neighbors that have maximum neighborhood influence on the protein of interest, then exclusion of non-essential neighbors may reduce the computation time. This is the basis of using heuristics adopted in FunPred-1.2. Using the heuristic that a higher neighborhood ratio may exist in densely connected sub-graphs the search space in FunPred-1.2 is reduced. Using the edge clustering coefficients, FunPred-1.2 finds the densely connected regions whereas edges with low edge clustering coefficients are not considered. In the reduced graph thus obtained similar procedure is followed up to predict functional group. The overall accuracy of FunPred-1.2 is 87% and average recall and precision are found to be $0.5613 \pm 0.1545$ and $0.8588 \pm 0.1192$. The F-score is $0.6724 \pm 0.1357$. Steps followed in FunPred-1 are described in details below.

## 4.1  FunPred 1.1

In FunPred 1.1 [21], the prediction technique is based on the combined score of neighborhood ratio, proteins path connectivity and relative functional similarity. Now, this method attempts to find the maximum of the summation of three scores thus obtained in each level and assign the unannotated protein to the corresponding functional group of the protein having the maximum value. Given $G'_p$, a sub graph of protein interaction network, consisting of proteins as nodes associated with any protein of set $O = \{O_1, O_2, O_3,...,O_8\}$; where, $O_i$ represents a particular functional group, this method maps the elements of the set of un-annotated proteins $P_U$ to any element of set O. Steps associated with this method are described below.

---

**Algorithm FunPred 1.1**

---

Input: Unannotated protein set $P_U$.
Output: The elements of the set of un-annotated
        proteins $P_U$ gets mapped to  any element of
        set O.

---

begin
   for all proteins in set $P_U$
      Count **Level** $- 1$ and **Level** $- 2$ neighbors of that
      protein in $G'_P$ associated with set O.
      Compute $P_{O_{i(=1,...,8)}}^{l(=1,2)}$ and assign this score to each
      protein $(\text{Pscore}^{l(=1,2)}) \in P_A$ belonging to the
      respective functional group.
      Compute $Q_{u,v}^{l(=1,2)}$, $W_{u,v}^{l(=1,2)}$ for each edge in **Level** $- 1$
      and **Level** $- 2$.
      Obtain neighborhood score i.e.
      $N_{(O_k)}^l = \text{Max}((\text{max}(\text{Pscore}^1 + Q_{u,v}^1 + W_{u,v}^1)), (\text{max}(\text{Pscore}^2 + Q_{u,v}^2 + W_{u,v}^2)))$
      Assign unannotated protein from the set $P_U$ to
      functional group $O_k$.
   end

---

## 4.2  FunPred 1.2

In FunPred 1.1, for any unannotated protein, we consider all Level-1 neighbors and
Level-2 neighbors belonging to any of 8 functional groups. Prediction is done on
the basis of neighborhood property where computation considers all Level-1 or
Level-2 neighbors. But if the computation is done only on significant neighbors
who have maximum neighborhood influence on the protein of interest then
exclusion of non-essential neighbors may reduce the time of computation. This is
the basis of our heuristic adopted in FunPred 1.2 [21]. Here, we only consider a
region or portion of a graph where neighbors are more connected, i.e., densely
connected neighbors are considered to be more significant. Steps associated with
this method are described below:

---

### Algorithm FunPred 1.2

---

Input:    Unannotated protein set $P_U$.
Output: The elements of the set of un-annotated
           proteins $P_U$ gets mapped to any element of
           set O.

---

begin
    for all proteins in set $P_U$
        Construct Neighborhood graph  $G'_P$ of the
        selected protein P with its **Level − 1** and **Level − 2**
        neighbors.
        Compute $ECC_{u,v}^{l(=1,2)}$ for each edge in **Level − 1** and
        **Level − 2**.
        Eliminate non-essential annotated proteins
        (neighbors) associated with edges having lower
        values of $ECC_{u,v}^{l(=1,2)}$ both in **Level − 1** and **Level − 2**.
        Update $G'_P$ with remaining nodes.
        Count **Level − 1**  and **Level − 2** neighbors of that
        protein in $G'_P$ associated with set O.
        Compute $P_{O_{i(=1,..8)}}^{l(=1,2)}$ and assign this score to each
        protein $(Pscore^{l(=1,2)}) \in P_A$ , belonging to the
        respective functional group.
        Compute  $Q_{u,v}^{l(=1,2)}$, $W_{u,v}^{l(=1,2)}$ for each edge in **Level − 1**
        and **Level − 2**.
        Obtain neighborhood score i.e.
        $N_{(O_k)}^l = Max((max(Pscore^1 + Q_{u,v}^1 + W_{u,v}^1 +$
                    $ECC_{u,v}^1)), (max(Pscore^2 + Q_{u,v}^2 + W_{u,v}^2 + ECC_{u,v}^2)))$
        Assign un-annotated protein from the set $P_U$ to
        functional group $O_k$.
    end.

---

Further it has been noted that analysis of the protein structure can provide functional clues or confirm tentative functional assignments inferred from the sequence. Many structure based approaches exist (e.g., fold similarity, three-dimensional templates), but as no single method can be expected to be successful in all cases, a more prudent approach involves combining multiple methods.

This observation tempted us to design our next work [20], a new approach to predict protein function is presented that combines sequential, structural information into protein-protein interaction network. Here physico-chemical properties of amino acids have been added in estimating neighborhood scoring values which help us to obtain more accuracy.

A protein may be involved in more than one activities resulting into interacting with different proteins of different functional groups. For this reason, selection of candidate protein (whose functional group is to be predicted) is very crucial task in this respect. If the candidate protein is orthodox in nature i.e., it is associated with a particular functional group then this method does not need to compute much more. On the other hand, it faces more challenges if the candidate protein is very popular individual (i.e., it interacts with protein of different functional groups) in protein interaction network. Locally dense regions are very likely to be protein complexes and proteins in the complex share the similar activity. So, candidate protein selection can be done from different protein complexes which ensure its popularity. In FunPred 2, given a protein interaction network, different dense regions or protein complexes are found using graph theoretic properties. The complex detection approach has been adopted from the work of Wang et al. [24]. From those complexes, candidate proteins are selected whose functional group is to be predicted. In every complex, thus found, of every protein $PCP_{score}^{l(=1,2)}$ is computed and $PCP_{score}^{l(=1,2)}$ of centroid of every complex is calculated. $PCP_{score}^{l(=1,2)}$ is defined as scaling of the mean value obtained from physico-chemical properties (Extinction Coefficient, Absorbance, Number of Negatively Charged Residues, Number of Positively Charged Residues, Aliphatic Index). For any candidate protein $PCP_{score}$ is calculated and using the similarity measure, it assigns candidate protein to a complex with which maximum similarity is found. Steps for finding Protein Complex are outlined below as *Algorithm A*:

---

### Algorithm A: **For finding Protein Complex**

```
Input:  Undirected simple PPI network G where every
        node represents a protein.
        Set threshold α and β for minimum edge weight
        and node weight respectively.
        Set m as number of clusters/complexes to be
        formed.
Output: Protein complexes/clusters
```

---

```
begin
    for all edges in G
        Compute edge weight
        if edge weight Wuv =0 then
          Remove corresponding edge.
    for all nodes in G
        Compute node weight
        if node weight does not exceed threshold β
          Remove corresponding node.
    Select a node K having highest node weight as a
    seed of cluster Cᵢ i.e. Cᵢ = {K}.
    Add neighbors of K to Cᵢ such that inclusion of
    neighbors of K does not cause edge  weight to
    fall below α i.e., Cᵢ = {K} ∪ Nₖ;  G = G − Cᵢ.
    Repeat the same procedure until m clusters are not
    formed.
end
```

---

Next, from each complex, 50% of proteins are taken randomly as test proteins and added to $P_U$, the set of non-annotated proteins whose functions have to be predicted. FunPred 2 predicts functional group of proteins of $P_U$ based on neighborhood analysis and physicochemical properties of their level-1 and level-2 proteins. For a particular protein in $P_U$, it selects significant level-1 neighbors (some level-1 neighbors having edge weight 0 and node weight less than β are deleted).For every significant level-1 neighbor of that protein, it finds their neighborhood graph comprising of their immediate neighbors using similar manner(considering the edge weight and node weight criterion). Thus, for a protein of interest having say, $q$ numbers of significant level-1 neighbors, q neighborhood graphs are formed. Next, it computes the $PCP_{score}$ of the protein of interest and also computes mean $PCP_{score}$ of its neighborhood graphs. Then, it computes difference between protein and each neighborhood graph's $PCP_{score}$. Finally, it assigns that protein to functional group of level-1 neighbor for which minimum difference is found. The following section describes the methodology of *FunPred 2*.

---

**Algorithm FunPred 2**

---

```
Input:    Set of un-annotated proteins P_U selected by
          Algorithm A
          Set threshold α for minimum edge weight.
Output:   Functional group of un-annotated proteins
```

---

```
begin
// Compute PCP score for every unknown proteins in P_u
   for all proteins in P_U
      Compute its PCP_score;
      for all level-1 neighbors
        Compute their node weights (W_v)
        Sort them according to their (W_v)
// construct neighborhood graph of every level-1
   neighbors
      for each level-1 neighbor K
        Make it a seed of cluster C_i i.e. C_i = {K}
        Add neighbors of K to C_i such that inclusion
        of neighbors of K does not cause edge weight to
        fall below α i.e., C_i = {K} ∪ N_K; G = G − C_i.
        for each cluster/neighborhood graph
          compute its mean PCP_score;
// assigning Functional Group to the proteins in P_u
      for each protein K in P_U
        for all clusters C_i
          find the difference of PCP_score of protein and
          mean PCP_score
      Assign the functional group of Cluster C_j to Protein
      K.
end
```

---

## 5  Results and Discussion

We have compared the performances of 4 methods: neighborhood counting method [1], the chi-square method [2], a recent version of the neighbor relativity coefficient (NRC) [23] and the FS-weight based method [25] for our *Saccharomyces cerevisiae* dataset with each other and with our previous methods as well as the current FunPred 2.

The NRC method [23], clearly performs better than the four methods. In Fig. 1, the performance of our earlier methods FunPred 1.1 and FunPred 1.2 across 8 functional groups highlights the fact that in terms of average prediction scores, our

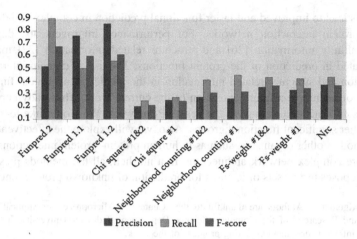

**Fig. 1** Comparative analysis existing methods with our developed method FunPred-1 and FunPred-2

method is better than the NRC method. This may be because both level-1 and level-2 neighbors are considered and a variety of scoring techniques in the protein interaction network, such as protein path connectivity, neighborhood ratio, and relative functional similarity etc. are also incorporated.

But it should be noted here that both FunPred 1.1 and FunPred 1.2 have low recall values and F-scores. The same is also true for the other methods. This has been completely surpassed by our current method FunPred 2. Though FunPred 2 has low Precision in comparison with that of FunPred 1.1 and FunPred 1.2 but it is subsequently higher than the remaining methods. This may be due to the fact that unannotated protein selection is independent of functional groups and subsequent incorporation of physicochemical properties of each protein in function prediction methodology. Moreover more number of functional groups are involved while choosing proteins randomly from each cluster. For chi-square methods (Chi-square#1 and Chi-square #1 and #2), the weak prediction has resulted due to the sparseness of PPI network as the chi-square methods work better on dense parts of the interaction network.

The result obtained in all Chi-square methods [2] is comparatively lower than the other methods because it only concentrates only on the denser region of the interaction network. The neighborhood counting method though performs well but fails when compared to NRC, FS-weight#1 (only direct neighbors are considered) and FS-weight #1 and #2 (both direct and indirect neighbors are considered) methods since it does not consider any difference between direct and indirect neighbors. Figure 1 shows a comparative detailed analysis of the four methods (taken into consideration in our work) along with our proposed systems.

Results show that our proposed FunPred 2 achieves better performance than existing function prediction methods. They also show that the network structure may be pruned based on the edge weight and along with it use of physico-chemical properties lead to improved and faster functional prediction in complex and diverse protein–protein interaction networks. For performance improvements, domain-domain affinity information [26] and structure related information [27] may be incorporated in prediction of the protein functions. The use of domain interaction information and structure related information in the prediction of protein function may be considered as a future extension of the current work. This study currently considers 18 functional groups in the yeast PPI network. We would like to extend this to other significant functional groups. Also, we will explore the effectiveness of this method in other organisms, such as in human protein–protein interactions with even more complex network architecture. In a nutshell, all the methods presented above proposes useful sets of features for prediction of unknown protein functions.

**Acknowledgements** Authors are thankful to the "Center for Microprocessor Application for Training and Research" of the Computer Science Department, Jadavpur University, India, for providing infrastructure facilities during progress of the work.

# References

1. Schwikowski, B., Uetz, P., Fields, S.: A network of protein-protein interactions in yeast. Nat. Biotechnol. **18**, 1257–1261 (2000)
2. Hishigaki, H., Nakai, K., Ono, T., Tanigami, A., Takagi, T.: Assessment of prediction accuracy of protein function from protein–protein interaction data. Yeast **18**, 523–31 (2001)
3. Chen, J., Hsu, W., Lee, M.L., Ng. S.K.: Labeling network motifs in protein interactomes for protein function prediction. In: IEEE 23rd International Conference on Data Engineering, pp. 546–555 (2007)
4. Vazquez, A., Flammini, A., Maritan, A., Vespignani, A.: Global protein function prediction from protein-protein interaction networks. Nat. Biotechnol. **21**, 697–700 (2003)
5. Karaoz, U., Murali, T.M., Letovsky, S., Zheng, Y., Ding, C., Cantor, C.R., Kasif, S.: Whole-genome annotation by using evidence integration in functional-linkage networks. Proc. Natl. Acad. Sci. **101**, 2888–2893 (2004)
6. Nabieva, E., Jim, K., Agarwal, A., Chazelle, B., Singh, M.: Whole-proteome prediction of protein function via graph-theoretic analysis of interaction maps. Bioinformatics **21**, i302–i310 (2005)
7. Deng, M., Mehta, S., Sun, F., Chen, T.: Inferring domain–domain interactions from protein–protein interactions. Genome. Res. 1540–1548 (2002)
8. Letovsky, S., Kasif, S.: Predicting protein function from protein/protein interaction data: a probabilistic approach. Bioinformatics **19**, i197–i204 (2003)
9. Wu, D. D.: An efficient approach to detect a protein community from a seed. In: IEEE Symposium on Computational Intelligence in Bioinformatics and Computational Biology, pp. 1–7 (2005)
10. Samanta, M.P., Liang, S.: Predicting protein functions from redundancies in large-scale protein interaction networks. Proc. Natl. Acad. Sci. **100**, 12579–12583 (2003)
11. Arnau, V., Mars, S., Marín, I.: Iterative cluster analysis of protein interaction data. Bioinformatics **21**, 364–378 (2005)
12. Bader, G.D., Hogue, C.W.V.: An automated method for finding molecular complexes in large protein interaction networks. BMC Bioinformatics **27**, 1–27 (2003)

13. Altaf-Ul-Amin, M., Shinbo, Y., Mihara, K., Kurokawa, K., Kanaya, S.: Development and implementation of an algorithm for detection of protein complexes in large interaction networks. BMC Bioinformatics **7**, (2006). doi:10.1186/1471-2105-7-207
14. Spirin, V., Mirny, L.A.: Protein complexes and functional modules in molecular networks. Proc. Natl. Acad. Sci. **100**, 12123–12128 (2003)
15. King, A.D., Przulj, N., Jurisica, I.: Protein complex prediction via cost-based clustering. Bioinformatics **20**, 3013–3020 (2004)
16. Asthana, S., King, O.D., Gibbons, F.D., Roth, F.P.: Predicting protein complex membership using probabilistic network reliability. Genome Res. **14**, 1170–1175 (2004)
17. Xiong, W., Liu, H., Guan, J., Zhou, S.: Protein function prediction by collective classification with explicit and implicit edges in protein-protein interaction networks. BMC Bioinform. **14**, Suppl 1, S4 (2013)
18. Zhao, B., Wang, J., Li, M., Li, X., Li, Y., Wu, F.X., Pan, Y.: A new method for predicting protein functions from dynamic weighted interactome networks. **15**, 131–139 (2016)
19. Saha, S., Chatterjee, P., Basu, S., Kundu, M., Nasipuri, M. (2012): Improving prediction of protein function from protein interaction network using intelligent neighborhood approach. In: IEEE International Conference on Communications, Devices and Intelligent Systems pp. 604–607
20. Saha, S., Chatterjee, P.: Protein function prediction from protein interaction network using physico-chemical properties of amino acid. Int. J. Pharm. Biol. Sci. **24**, 55–65 (2014)
21. Saha, S., Chatterjee, P., Basu, S., Kundu, M., Nasipuri, M.: Funpred-1: Protein Function Prediction From A Protein Interaction Network Using Neighborhood Analysis. Cell. Mol. Biol. Lett. (2014). doi:10.2478/s11658-014-0221-5
22. Wu, X., Zhu, L., Guo, J., Zhang, D.Y., Lin, K.: Prediction of yeast protein-protein interaction network: insights from the Gene Ontology and annotations. Nucleic Acids Res. **34**, 2137–2150 (2006)
23. Moosavi, S., Rahgozar, M., Rahimi, A.: Protein function prediction using neighbor relativity in protein-protein interaction network. Comput. Biol. Chem. **43** (2013). doi:10.1016/j.compbiolchem.2012.12.003
24. Wang, S., Wu, F.: Detecting overlapping protein complexes in PPI networks based on robustness. Proteome Sci. **11**, S18 (2013)
25. Chua, H.N., Sung, W.K., Wong, L.: Exploiting indirect neighbours and topological weight to predict protein function from protein-protein interactions. Bioinformatics **22**, 1623–1630 (2006)
26. Chatterjee, P., Basu, S., Kundu, M., Nasipuri, M., Plewczynski, D.: PSP_MCSVM: brainstorming consensus prediction of protein secondary structures using two-stage multiclass support vector machines. J. Mol. Model. **17**, 2191–2201 (2011)
27. Chatterjee, P., Basu, S., Zubek, J., Kundu, M., Nasipuri, M., Plewczynski, D.: PDP-CON: prediction of domain/linker residues in protein sequences using a consensus approach. J. Mol. Model. (2016). doi:10.1007/s00894-016-2933-0

# Author Index

© Springer Nature Singapore Pte Ltd. 2017
R. Chaki et al. (eds.), *Advanced Computing and Systems for Security*,
Advances in Intelligent Systems and Computing 568,
DOI 10.1007/978-981-10-3391-9

Printed in the United States
By Bookmasters